# Journal of Applied Logics - IfCoLog Journal of Logics and their Applications

Volume 7, Number 5

September 2020

**Disclaimer**
Statements of fact and opinion in the articles in Journal of Applied Logics - IfCoLog Journal of Logics and their Applications (JALs-FLAP) are those of the respective authors and contributors and not of the JALs-FLAP. Neither College Publications nor the JALs-FLAP make any representation, express or implied, in respect of the accuracy of the material in this journal and cannot accept any legal responsibility or liability for any errors or omissions that may be made. The reader should make his/her own evaluation as to the appropriateness or otherwise of any experimental technique described.

© Individual authors and College Publications 2020
All rights reserved.

ISBN 978-1-84890-345-6
ISSN (E) 2631-9829
ISSN (P) 2631-9810

College Publications
Scientific Director: Dov Gabbay
Managing Director: Jane Spurr

http://www.collegepublications.co.uk

---

All rights reserved. No part of this publication may be reproduced, stored in a retrieval system or transmitted in any form, or by any means, electronic, mechanical, photocopying, recording or otherwise without prior permission, in writing, from the publisher.

# Editorial Board

Editors-in-Chief
Dov M. Gabbay and Jörg Siekmann

Marcello D'Agostino
Natasha Alechina
Sandra Alves
Arnon Avron
Jan Broersen
Martin Caminada
Balder ten Cate
Agata Ciabttoni
Robin Cooper
Luis Farinas del Cerro
Esther David
Didier Dubois
PM Dung
David Fernandez Duque
Jan van Eijck
Marcelo Falappa
Amy Felty
Eduaro Fermé

Melvin Fitting
Michael Gabbay
Murdoch Gabbay
Thomas F. Gordon
Wesley H. Holliday
Sara Kalvala
Shalom Lappin
Beishui Liao
David Makinson
George Metcalfe
Claudia Nalon
Valeria de Paiva
Jeff Paris
David Pearce
Pavlos Peppas
Brigitte Pientka
Elaine Pimentel

Henri Prade
David Pym
Ruy de Queiroz
Ram Ramanujam
Chrtian Retoré
Ulrike Sattler
Jörg Siekmann
Jane Spurr
Kaile Su
Leon van der Torre
Yde Venema
Rineke Verbrugge
Heinrich Wansing
Jef Wijsen
John Woods
Michael Wooldridge
Anna Zamansky

# Scope and Submissions

This journal considers submission in all areas of pure and applied logic, including:

| | |
|---|---|
| pure logical systems | dynamic logic |
| proof theory | quantum logic |
| constructive logic | algebraic logic |
| categorical logic | logic and cognition |
| modal and temporal logic | probabilistic logic |
| model theory | logic and networks |
| recursion theory | neuro-logical systems |
| type theory | complexity |
| nominal theory | argumentation theory |
| nonclassical logics | logic and computation |
| nonmonotonic logic | logic and language |
| numerical and uncertainty reasoning | logic engineering |
| logic and AI | knowledge-based systems |
| foundations of logic programming | automated reasoning |
| belief change/revision | knowledge representation |
| systems of knowledge and belief | logic in hardware and VLSI |
| logics and semantics of programming | natural language |
| specification and verification | concurrent computation |
| agent theory | planning |
| databases | |

This journal will also consider papers on the application of logic in other subject areas: philosophy, cognitive science, physics etc. provided they have some formal content.

Submissions should be sent to Jane Spurr (jane@janespurr.net) as a pdf file, preferably compiled in LaTeX using the IFCoLog class file.

# CONTENTS

## ARTICLES

Semantic Spaces at the Intersection of NLP, Physics,
    and Cognitive Science . . . . . . . . . . . . . . . . . . . . . . . . . 677
   *Martha Lewis, Dan Marsden and Mehrnoosh Sadrzadeh*

Integrating Conceptual Spaces in Frames . . . . . . . . . . . . . . . . . . 683
   *Corina Strößner*

A Vector Simplex Model of Concepts . . . . . . . . . . . . . . . . . . . 707
   *Douglas Summers-Stay*

Concept Functionals . . . . . . . . . . . . . . . . . . . . . . . . . . . . . 725
   *Vincent Wang*

Meaning Updating of Density Matrices . . . . . . . . . . . . . . . . . . 745
   *Bob Coecke and Konstantinos Meichanetzidis*

Towards Logical Negation for Compositional Distributional Semantics . . . 771
   *Martha Lewis*

Density Matrices with Metric for Derivational Ambiguity . . . . . . . . . . . 795
   *Adriana D. Correia, Michael Moortgat and Henk T. C. Stoof*

A Frobenius Algebraic Analysis for Parasitic Gaps . . . . . . . . . . . . . 823
   *Michael Moortgat, Mehrnoosh Sadrzaden and Gijs Wijnholds*

**Vector Spaces as Kripke Frames** . . . . . . . . . . . . . . . . . . . . . . . . . . . 853
*Giuseppe Greco, Fei Liang, Michael Moortgat, Alessandra Palmigiano and Apostolos Tzimoulis*

# Semantic Spaces at the Intersection of NLP, Physics, and Cognitive Science

Martha Lewis*
*Dept. of Engineering Mathematics, University of Bristol*
*ILLC, University of Amsterdam*
martha.lewis@bristol.ac.uk

Dan Marsden
*Dept. of Computer Science, University of Oxford*
daniel.marsden@cs.ox.ac.uk

Mehrnoosh Sadrzadeh
*Dept. of Computer Science, University College London*
m.sadrzadeh@ucl.ac.uk

## Abstract

The ability to compose parts to form a more complex whole, and to analyze a whole as a combination of elements, is desirable across disciplines. Semantic Spaces at the Intersection of Natural Language Processing (NLP), Physics, and Cognitive Science brought together researchers applying similar compositional approaches within the three disciplines. The categorical model of [6], inspired by quantum protocols, has provided a convincing account of compositionality in vector space models of NLP. Similar category-theoretic approaches have been applied in cognitive science, in the context of conceptual spaces. The interplay between the three disciplines fostered theoretically motivated approaches to understanding how meanings of words interact in sentences and discourse, and how concepts develop in a cognitive space. This volume sees commonalities between the compositional mechanisms employed extracted, and applications and phenomena traditionally thought of as 'non-compositional' being shown to be compositional.

Many thanks to the programme committee for their hard work reviewing the papers that contributed to this volume, and to the organisers of ESSLLI 2019 for providing a venue that made the event such a success.

*Funded by NWO Veni project 'Metaphorical Meanings for Artificial Agents'

# 1 Background

Since their introduction in the early 1970s, vector space models of meaning have evolved into a well-established area of research in NLP. Their probabilistic nature and ability to exploit the abundance of large-scale resources such as the Web make them one of the most useful tools (arguably the most successful [23]) for modelling what we broadly call *meaning* in language.

There is also a long-standing history of vector space models in cognitive science. Theories of categorization such as those developed by [1, 14, 18] utilise notions of distance and similarity that can readily be incorporated in vector space models of meaning. [12, 19, 24] encode meanings as feature vectors, and models of high-level cognitive reasoning have been implemented within vector symbolic architectures [16, 20, 11]. More recently [8, 9] has developed a model of concepts in which *conceptual spaces* provide geometric structures, and information is represented by points, vectors and regions in vector spaces. The conceptual spaces model has been applied to language evolution [22], scientific theory change [10], and models of musical creativity [7], amongst others, and has the potential to augment NLP models of meaning with representations that have been learned through interaction with the external world.

A third field in which vector space models play an important role is physics, and especially quantum theory. Though seemingly unrelated to language, intriguing connections have recently been uncovered. The link between physics and natural language semantics that vector space models provide has been successfully exploited, providing novel solutions and a fresh perspective for a number of problems related to NLP and cognitive science, such as modelling logical aspects in vector spaces [26]. Methods from quantum logic have also been applied to cognitive processes related to the human mental lexicon, such as word association [4], decision-making [17], human probability judgements [5], and information retrieval [25]. Furthermore, the categorical model of [6], inspired by quantum mechanics, has provided a convincing account of compositionality in vector space models and an extensible framework for linguistically motivated research on sentential semantics. More recently, the link between physics and text meaning was made more concrete by a number of proposals that aim at replacing the traditional notion of a word vector with that of a density matrix—a concept borrowed from quantum mechanics which can be seen as a probability distribution over vectors [15, 3, 21].

## 2  Topics covered

Exploiting the common ground provided by semantic spaces, the SemSpace workshop brought together researchers working at the intersection of NLP, cognitive science, and physics, offering to them an appropriate forum for presenting their uniquely motivated work and ideas. The workshop attracted 15 submissions, of which 8 have been developed into full length papers for this journal issue.

Three papers covered the representation of concepts in semantic spaces. *Integrating Conceptual Spaces in Frames* describes how the theories of conceptual spaces [8] and frames [2] can be brought together in a single theory that addresses their respective limitations. *A Vector Simplex Model of Concepts* lays out a vector-based memory and reasoning system that aims to unify the main theories of concepts: *classical*, *prototype*, *exemplar*, and the *theory-theory*. *Concept Functionals* brings a more compositional aspect to the representation of concepts. Concepts are modelled within spaces of functionals that form a *compact closed category*, meaning that they can be the target of a functor from pregroup grammar [13] that describes how words compose according to grammatical structure, as in the categorical compositional distributional (DisCoCat) model of meaning [6].

Three papers use the concept of a *density matrix* to represent words and phrases. Density matrices allow the representation of a probability distribution over vectors, and can also be incorporated into a categorical compositional account. In *Meaning Updating for Density Matrices* an extension of DisCoCat is used under the acronym DisCoCirc. This extension starts to show how larger fragments of text can be be composed to form narratives. The paper describes two update mechanisms for density matrices and shows how the two update mechanisms can be combined in a shared categorical representation. In *Towards Logical Negation for Compositional Distributional Semantics* a notion of logical negation is introduced that is akin to projection to the orthogonal subspace of a vector, and this notion of negation, together with various composition operators, is evaluated on a short entailment dataset. *Density Matrices with Metric for Derivational Ambiguity* use density matrices to represent ambiguity in grammatical parses, rather than in semantic content. To do so, the authors use a variant of Lambek's categorial grammar with directional implication, and build a canonical isomorphism between a vector space and its dual that allows the directional implication to be retained, unlike in standard pregroup grammar.

Advanced linguistic structure is examined in *A Frobenius Algebraic Analysis for Parasitic Gaps*. Here, the notion of a gap being felicitous on the presence of another gap, is modelled. On the syntactic side, Lambek calculus with structural control modalities are used. On the semantic side, Frobenius algebras are employed. The reliance on the over-generating operations of copying and moving, often used to model parasitic gaps, is overcome by the novel use of Frobenius algebras, from quantum mechanics, as a vessel for delivering type polymorphism. Finally, *Vector Spaces as Kripke Frames* extends the DisCoCat approach from focussing on pregroup grammar to a vector space semantics for the general Lambek calculus, based on algebras over a field, meaning that the match between the grammar of natural languages such as English can be much more closely matched. Here, general ordered algebraic operations are used to model different operations on vector spaces, opening up the field for operations that are not necessarily associative to commutative.

We also had three invited speakers: Ruth Kempson, Jamie Kiros, and Sanjaye Ramgoolam. Ruth Kempson described the Dynamic Syntax view that a natural language grammar is a set of processes inducing incremental context-relative coordination of action. Jamie Kiros argued for the importance of research in language grounding, structural priors and their relevance to constructing semantic spaces, particularly in the light of recent advances in large-scale language modelling and contextualized word representations that have led to significant improvements across several language processing tasks. Sanjaye Ramgoolam presented ideas from random matrix theory in physics that have been applied to characterize the statistics of matrices for adjectives and verbs generated in compositional distributional semantics. Permutation invariance was argued to be the appropriate symmetry.

The range of topics covered allowed lively discussion and ideas around theoretically motivated approaches to understanding how meanings of words interact with each other in sentences and discourse, how they are determined by input from the world, and how word and sentence meanings interact logically.

## Acknowledgements

Semantic Spaces at the Intersection of NLP, Physics, and Cognitive Science was supported by NWO Veni project 'Metaphorical Meanings for Artificial Agents'.

## References

[1] F Gregory Ashby and Ralph E Gott. Decision rules in the perception and categorization of multidimensional stimuli. *Journal of Experimental Psychology: Learning, Memory, and Cognition*, 14(1):33, 1988.

[2] Lawrence W Barsalou. Frames, concepts, and conceptual fields. In A. Lehrer & E. F. Kittay, editor, *Frames, fields, and contrasts: New essays in semantic and lexical organization*, pages 21–74. Lawrence Erlbaum Associates, Inc, 1992.

[3] William Blacoe, Elham Kashefi, and Mirella Lapata. A quantum-theoretic approach to distributional semantics. In *Proceedings of the 2013 Conference of the North Ameri-*

can *Chapter of the Association for Computational Linguistics: Human Language Technologies*, pages 847–857, Atlanta, Georgia, June 2013. Association for Computational Linguistics.

[4] Peter Bruza, Kirsty Kitto, Douglas Nelson, and Cathy McEvoy. Is there something quantum-like about the human mental lexicon? *Journal of Mathematical Psychology*, 53(5):362–377, 2009.

[5] Jerome R Busemeyer, Emmanuel M Pothos, Riccardo Franco, and Jennifer S Trueblood. A quantum theoretical explanation for probability judgment errors. *Psychological Review*, 118(2):193, 2011.

[6] Bob Coecke, Mehrnoosh Sadrzadeh, and Stephen Clark. Mathematical Foundations for a Compositional Distributional Model of Meaning. Lambek Festschrift. *Linguistic Analysis*, 36:345–384, 2010.

[7] Jamie Forth, Geraint A Wiggins, and Alex McLean. Unifying conceptual spaces: Concept formation in musical creative systems. *Minds and Machines*, 20(4):503–532, 2010.

[8] Peter Gärdenfors. *Conceptual spaces: The geometry of thought*. MIT press, 2004.

[9] Peter Gärdenfors. *The geometry of meaning: Semantics based on conceptual spaces*. MIT Press, 2014.

[10] Peter Gärdenfors and Frank Zenker. Theory change as dimensional change: Conceptual spaces applied to the dynamics of empirical theories. *Synthese*, 190(6):1039–1058, 2013.

[11] Ross W Gayler. Vector symbolic architectures answer Jackendoff's challenges for cognitive neuroscience. *arXiv preprint cs/0412059*, 2004.

[12] James A Hampton. Inheritance of attributes in natural concept conjunctions. *Memory & Cognition*, 15(1):55–71, 1987.

[13] Joachim Lambek. Type grammars as pregroups. *Grammars*, 4(1):21–39, 2001.

[14] Robert M Nosofsky. Attention, similarity, and the identification–categorization relationship. *Journal of Experimental Psychology: General*, 115(1):39, 1986.

[15] Robin Piedeleu, Dimitri Kartsaklis, Bob Coecke, and Mehrnoosh Sadrzadeh. Open System Categorical Quantum Semantics in Natural Language Processing. In *Proceedings of the 6th Conference on Algebra and Coalgebra in Computer Science (CALCO)*, Nijmegen, Netherlands, June 2015.

[16] Tony A. Plate. Holographic Reduced Representations: Convolution Algebra for Compositional Distributed Representations. In *Proceedings of the 12th International Joint Conference on Artificial Intelligence*, pages 30–35. Citeseer, 1991.

[17] Emmanuel M Pothos and Jerome R Busemeyer. Can quantum probability provide a new direction for cognitive modeling? *Behavioral and Brain Sciences*, 36(03):255–274, 2013.

[18] Eleanor Rosch and Carolyn B Mervis. Family resemblances: Studies in the internal structure of categories. *Cognitive Psychology*, 7(4):573–605, 1975.

[19] Edward E Smith and Daniel N Osherson. Conceptual combination with prototype concepts. *Cognitive Science*, 8(4):337–361, 1984.

[20] Paul Smolensky. Tensor Product Variable Binding and the Representation of Symbolic

Structures in Connectionist Systems. *Artificial Intelligence*, 46:159–216, 1990.

[21] Alessandro Sordoni and Jian-Yun Nie. Looking at vector space and language models for ir using density matrices. In *Quantum Interaction*, pages 147–159. Springer, 2014.

[22] Luc Steels and Tony Belpaeme. Coordinating perceptually grounded categories through language: A case study for colour. *Behavioral and brain sciences*, 28(4):469–488, 2005.

[23] Peter D Turney and Patrick Pantel. From frequency to meaning: Vector space models of semantics. *Journal of artificial intelligence research*, 37(1):141–188, 2010.

[24] Amos Tversky. Features of similarity. *Psychological Review*, 84(4):327–352, 1977.

[25] Cornelis Joost Van Rijsbergen. *The geometry of information retrieval*, volume 157. Cambridge University Press Cambridge, 2004.

[26] Dominic Widdows. Orthogonal negation in vector spaces for modelling word-meanings and document retrieval. In *Proceedings of the 41st Annual Meeting on Association for Computational Linguistics-Volume 1*, pages 136–143. Association for Computational Linguistics, 2003.

# Integrating Conceptual Spaces in Frames

Corina Strößner*
*Department of Philosophy II, Ruhr University Bochum, Universitätstrasse 150,
44780 Bochum, Germany*
corina.stroessner@ruhr-uni-bochum.de

## Abstract

Conceptual spaces and recursive frames are two different models of conceptual representation. This paper explains their commonalities and differences. We argue that these seemingly competing methods can complement each other. Frames can relate conceptual spaces in different domains making them an attractive supplement for the theory of conceptual spaces. In contrast, frames benefit from a more fine-grained and structured representation of properties that can be only provided by conceptual spaces. Based on existing overlaps in research, the paper explores how combinations of frames and conceptual spaces can be generated and how they can be used to develop a better understanding of natural concepts.

## 1 Conceptual Content and Its Representation

Many theories claim that concepts are determined by cognitive content. This position is called *cognitivism*. Cognitivism is to be distinguished from approaches that understand concepts primarily as denoting entities in the world (e.g. [9]). Cognitivist positions need to address two questions. First, *what* is conceptual content? Second, *how* is it represented? There are many answers to the first question. Concepts were traditionally considered to be definitions. However, over the last decades, other accounts became more prominent. Eleanor Rosch and her collaborators [29, 31, 32, 30] regard concepts as prototype structures that capture similarities between category members. This is known as the *prototype theory*. Another position, known as *theory*

---

I like to thank the participants of the SemSpace 2019 workshop and especially the organizers of the event. Moreover I am grateful for helpful comments from Martha Lewis and three anonymous reviewers. The ideas in this paper were influenced by many fruitful discussions with Gerhard Schurz, Annika Schuster, Peter Sutton and Henk Zeevat.

*This work was generously supported by the German Research Foundation (DFG), grant SFB991/D01.

*theory*, analyses concepts in terms of their role in world knowledge and background theories (cf. [25]). This paper does not commit itself to any of these theories. Each of these approaches has its justification depending on the concepts and domain involved. For example, mathematical and everyday concepts differ fundamentally and need different theories of concepts.[1] In this paper, concepts are broadly understood as representations of categories. These are collections of possible entities, which will be called members of the category.

There are currently two highly researched and frequently applied models of conceptual representation: conceptual spaces and recursive frames. Both aim to structure conceptual content. Proponents of the two approaches conduct their research independently, but they address similar issues in linguistics, cognitive science, and philosophy. While proponents of conceptual spaces and frame theory address comparable questions, they disagree on the appropriate representational scheme. This motivates the first aim of this paper: a discussion of the commonalities and differences as well as the respective advantages and limitations of frames and conceptual spaces. Building on this comparison, the second aim will be to discuss possible integrations. The paper proceeds with a survey of frames and conceptual spaces in section 2 and 3. Section 4 gives a comparison between the models. The integration of conceptual spaces in frames is discussed in section 5 and 6.

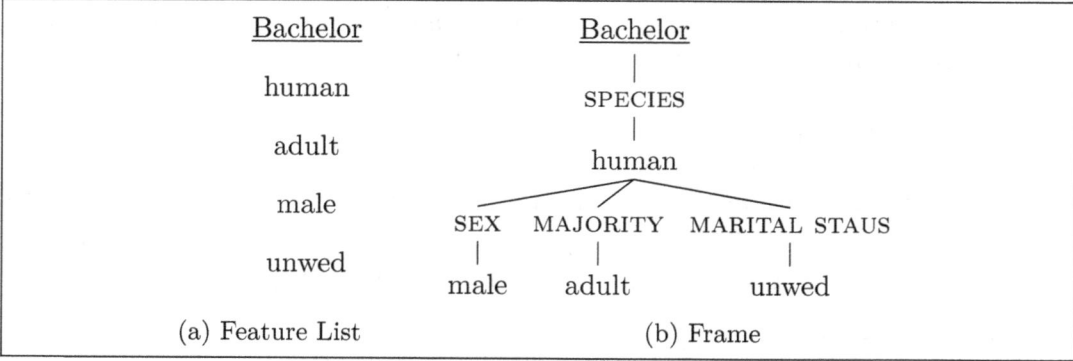

Figure 1: Representation of the meaning of 'bachelor' (defined as unmarried man) in a feature list (fig. 1a) and in a frame (fig. 1b).

## 2 Frames

Frames originated in Minsky [24] and were further developed in the field of cognitive science by Barsalou [3]. They extend feature lists by a functional structure, which consists of attributes and values. For example, a bachelor is not represented as an unmarried male but as a human with the attributes SEX and MARITAL STATUS and their values 'male' and 'unwed'.[2] This attribute value structure is recursive, that is, it is nested: a value of one attribute can be the argument of another. Figure 1 illustrates how the attributes of MARITAL STATUS, MAJORITY and SEX are applied to a value of another attribute: 'human'.

In addition to the recursive attribute value structure, Barsalou [3] discusses relationships between attributes and values. *Structural invariants* are relations between attributes. For example, concepts with the attribute SENDER, like 'information' or 'message', should also incorporate RECEIVER. *Constraints* state dependencies between the *values* of different attributes. For example, the values of MARITAL STATUS and AGE are related: younger people are more likely to be unwed than older ones.

**Graph Theoretic Definition of Frames in Linguistics**  Petersen [27] developed a precise formalisation of recursive frames in terms of graph theory, which is widely used in linguistics [19, 23]. According to her definition, frames consist of a set of finite nodes $V$. One is distinguished as the central node $v$ that stands for the represented concept. The nodes are connected by a partial attributing function $att$ that assigns one node to a tuple of nodes, together with an attribute label from the set $A$. The nodes gain their meanings by a function $typ$ that assigns a type from the set of types $T$ to them. The type specifies appropriate values. For example, the value for AGE is necessarily a time, and for humans it is a value between 0 and approximately 120 years. Given the tuple $\langle V, v, T, typ, A, att \rangle$, the directed graph $\langle V, \vec{E} \rangle$, where $\vec{E}$ are nodes connected by the attributing function, is the frame.

Often the root of a frame is also its central node. For example, in the 'bachelor' frame in figure 1, the represented concept 'bachelor' is the one to which all further attributes are applied, either directly or via other values. Concepts with such a frame structure are *sortal concepts*. However, frames are by no means limited to such structures. For example, the *functional concept* 'mother' appears as a value of the according attribute MOTHER. Another kind of frames are found for *relational concepts* like 'brother': they have two roots. This and further examples from [27] are provided in figure 2.

---

[1] Arguments for conceptual pluralism can be inter alia found in [33]

[2] We use capitals for attributes. For other concepts, we use single quotation marks. Double quotation marks are used for quotes, objective language, and non-literal speech.

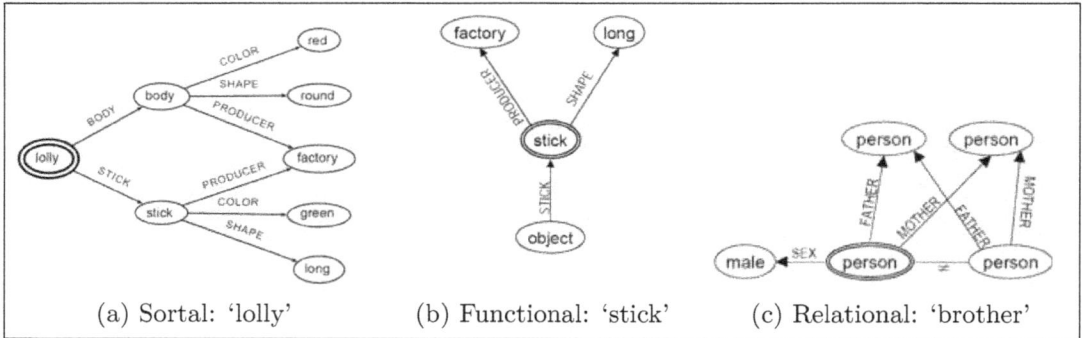

Figure 2: Exemplary graphs, directly taken from [27, p. 158]: The central node is the root for sortal concepts (fig. 2a) but not for functional concepts (fig. 2b). Relational concepts (fig. 2c) have more than one root, one of which represents the concept.

With their increased level of formality and generality, Petersen's frames are useful in natural language processing. While they ignore several aspects of Barsalou's model (most notably knowledge constraints), the two main points remain central. First, attributes are functional: only one node can be its value; second, nodes occur as arguments of attributes as well as their values. Both points together yield the recursive attribute value structure of frames.

**Classification Frames in Philosophy of Science** Independently from the linguistic formalisation, philosophers of science adopted frames to represent the development of scientific theories and theoretical concepts [7, 2, 38, 21, 20]. In this paper they are called *classification frames*.

Classification frames have a root that represents the application domain of a theory, expressed in terms of a category. Several attributes are applied to it and alternative values are represented. Constraints, that is dependencies between values of different attributes, model correlations. Subordinated categories are derived from combinations of values. Figure 3 shows an example from [20]. It represents Perlmutter's distinction between pro drop and non-pro drop languages [26] by three grammatical characteristics, namely the possibility to drop thematic and non-thematic subjects and to extract the subject of a subordinate clause. The dashed lines are constraints. They illustrate the relation between these properties. The subcategories capture these constraints: pro drop languages realise all grammatical parameters while non-pro drop languages have none of them.

This version of frames differs from the one of Petersen [27] in several aspects. First, classification frames are more informative since they elucidate different al-

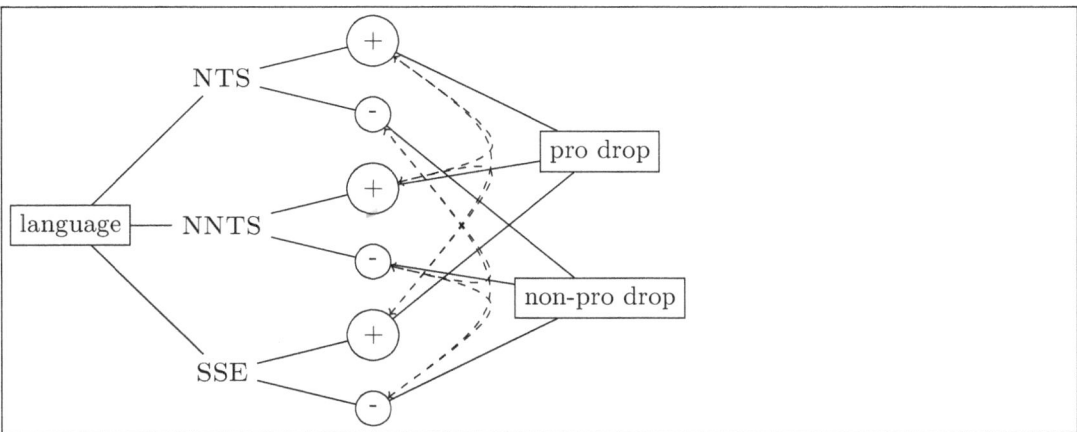

Figure 3: Frame from [20] for representing the distinction between pro and non-pro language made by [26]. In pro drop languages it is permissible to omit subject pronouns (NTS), non-thematic subjects (NNTS) and to extract the subject of a subordinate clause (SSE) [20, p. 151-52]

ternative values and represent constraints between them. On the other hand, they have a quite fixed structure and are thus less flexible. However, classification frames and the graph-theoretic frames from [27] share core assumptions from Barsalou [3]. They represent categories in terms of attributes of which every category member instantiates exactly one possible value and they are recursive. Though recursivity is restricted by the fixed structure in classification frames, subcategories can be analysed in terms of further frame structures, rendering them also recursive. For example, pro-drop languages could be further analysed in terms of other grammatical parameters.

To conclude, frames are essentially recursive attribute value structures. From this common ground, several versions of frames were developed, which all agree on this central point.

## 3 Conceptual Spaces

The fundamental ideas concerning conceptual spaces were presented in the seminal book "Conceptual spaces: The geometry of thought" by Gärdenfors [10] within which he claims that geometrical methods provide the appropriate tool for conceptual representation. According to him, a vast number of basic concepts, most importantly the adjectives from perceptual domains, can be represented as areas in a space. The most popular example is the colour space.

There are several formally detailed versions of conceptual spaces. Aisbett and Gibbon [1] define conceptual spaces in pointed metric spaces, that is, sets with a distance function $d$ and a point of infinity, which is maximally distant from all other points. Raubal [28] uses vectors to define his conceptual spaces. More recent approaches combine conceptual spaces with random set theory [22] or fuzzy set theory [5]. All these approaches refer to Gärdenfors [10] and his understanding of conceptual spaces:

> A conceptual space consists of a class $D_1, \ldots, D_n$ of quality dimensions. A *point* in the space is represented by a vector $v = \langle d_1, \ldots, d_n \rangle$, with one index for each dimension. Each dimension is endowed with a certain geometrical or topological structure [10, p. 67].

Though it is not excluded that the dimensions of a conceptual space are merely binary or comparative, most discussions on conceptual spaces concern metric spaces. These are pairs $(X, d)$ consisting of a set $X$ and a real-valued distance measure $d$ over pairs of elements from $X$ such that $d$ satisfies the following axioms:

D1 $d(x,y) \geq 0$; $d(x,y) = 0$ if and only if $x = y$ (minimality).

D2 $d(x,y) = d(y,x)$ (symmetry).

D3 $d(x,z) + d(y,z) \geq d(x,y)$ (triangle inequality).
 cf. [10, p. 18]

The distance between two points in the space depends on their distance in the single dimensions. Gärdenfors [10, p. 20] presents two measures to determine the distance between points in multidimensional spaces with each dimension being accompanied by an importance weight $w_i$:

Euclidean distance: $d(x,y) = \sqrt{\sum_{i=1}^{n} w_i (x_i - y_i)^2}$,

City-block (Manhattan) distance: $d(x,y) = \sum_{i=1}^{n} w_i |x_i - y_i|$.

Both are special cases of the more general

Minkowski distance: $d(x,y) = (\sum_{i=1}^{n} w_i |x_i - y_i|^p)^{1/p}$.[3]

---

[3]There are other ways to measure similarity, e.g., cosine similarity but Minkowski distances dominate the discussion of conceptual spaces.

From the Minkowski distance, the city-block metric is defined by $p = 1$ and the Euclidean distance by $p = 2$. There are conceptual implications associated with the particular value of $p$. Euclidean distances are appropriate in spaces with integral dimensions, that is, dimensions that are so deeply connected that, by assigning a value to one dimension, one also has to assign values to the other dimensions.[4] A bundle of integral dimensions is called *domain*.

Since the Minkowski distance allows gradations between Euclidean and Manhattan distances, there can be degrees of integrality. Moreover, as discussed by [18], different pairs of dimensions in a space can have different measures. As an example, think of a conceptual space of coloured, differently sized dots. It can be expected that the colour specific dimensions HUE, SATURATION and BRIGHTNESS combine in a Euclidean manner, yet these dimensions have no Euclidean distance to the dimension of SIZE.[5]

A central idea in the theory of conceptual spaces is that (natural) concepts are topologically restricted. The most popular principle is convexity. It relies on the notion of betweenness $B(a, b, c)$, read as $b$ is between $a$ and $c$. An area is convex if and only if all points between any two points of the region are also in the region. Convexity is thus closure under betweenness.

Most frequently convexity is applied in spaces with Euclidean metrics, that is, domains. Gärdenfors [10, p. 71] specifies a *natural property* as a convex region in a domain. On other occasions, Gärdenfors suggests convexity as general criterion of natural concepts without the restriction to properties: "A *natural concept* is a convex region of a conceptual space" [11, p. 18]. There is significant evidence for convexity in domains, that is, Euclidean spaces. Using evolutionary game theory, Jäger [17] demonstrates that, given plausible background assumptions, language evolution divides domains into convex regions. In addition, neighbourhood-matching methods (e.g., Voronoi tesselation) yield convex regions in Euclidean spaces. Convexity in other conceptual spaces is more problematic. A critical view of convexity in non-integral dimensions is taken by Hernández-Conde [16]. Recently, Gärdenfors [12] clarified that he views convexity as an empirical thesis and not as an analytically true assumption of conceptual spaces theory. As such, it can be confirmed or refuted by further research. Moreover, many proponents of conceptual spaces do not suppose it at all [5]. The present investigation thus views conceptual spaces theory as being committed to representing conceptual content in terms of topology and geometry but not to the convexity thesis.

---

[4]The term "dimension" is slightly ambiguous as it can mean a function or its value space. We will use it as a counterpart of attributes in frames (i.e., as function).

[5]This is an example for explanatory purposes. Johannesson [18] outlines more complex experimental data.

## 4 Comparison

As argued by Barsalou and Hale [4], frames clearly extend the representational possibilities of simple feature lists. However, there is no consensus on whether conceptual spaces exceed the power of frames. Zenker [40] argues that the important components of frames are recoverable in conceptual spaces. Furthermore, only the latter address quantitative (interval, ratio, or absolute-scaled) measurement scales, which are alien to frames. This claim was rejected by Votsis and Schurz [38]. The question of how frames and conceptual spaces are related to each other is thus not settled and it is also barely researched. Zenker [40] placed his brief comparison in a discussion of scientific change and it can be doubted that he primarily intended a thorough comparison of frames and conceptual spaces. Votsis and Schurz [38] answer in barely more than a footnote.[6]

This section aims to offer a comparative overview of the commonalities and differences between the two approaches, based on their respective cores: the modelling of concepts in terms of recursive attribute value structures in frames and the geometric viewpoint on concepts, following the tradition of Gärdenfors [10].

Frames and conceptual spaces both assume that concepts should be analysed in terms of functions and values. Frame theorists call these "attributes". Regarding conceptual spaces, one uses "dimensions", emphasising a geometric structure of the value spaces. By doing so, both frameworks promote a distinction between functions and values. A single instance of a concept always takes one specific value, for example, one point in the conceptual spaces. This fundamental commonality makes it important for both approaches to discuss which attributes or dimensions contribute to the conceptual representation and how important they are. For the frames in Petersen [27], the bare "skeleton" of attributes with their range and value spaces does most of the representational work. In conceptual spaces, the question of contributing dimensions (and domains) is fundamental as well. For example, Gärdenfors [12] emphasises that SHAPE rather than COLOUR is critical in the representation of 'swan'.

The most apparent difference between frames and conceptual spaces is the role of quantitative and nominal values. As previously stated, Zenker [40] claims that conceptual spaces extend frames because they allow for quantitatively scaled value spaces. Votsis and Schurz [38] object that a frame "allows values to be structured in terms of nominal, ordinal, interval, ratio or absolute scales" [38, p. 108]. Kornmesser and Schurz [21] actually use a quantitative classification frame of electrostatics in which constraints are given as equations, namely versions of Coulomb's law.[7]

---

[6] In a personal communication, Frank Zenker as well as Gerhard Schurz elaborated their position, which became a main inspiration for the present investigation.

[7] The frame and a detailed explanation of it is found in section 4 of their paper.

On closer inspection, this dispute can be dissolved by distinguishing two questions: *what* kind of information can be represented and *how* is it represented? With respect to the former question, frames are indeed not restricted in any way. All kinds of attributes are representable, as Votsis and Schurz [38] point out and as Kornmesser and Schurz [21] actually demonstrate by an example. Though discrete (i.e., nominal) values are more common in frames, they are not limited to them. Conceptual spaces, on the other hand, are usually applied to quantitative scales, but ordinal and even merely classificatory values are not excluded [6]. Differences between frames and conceptual spaces in their ability to represent different measurement scales are thus only tendentious.

With respect to the second question, however, there is a true difference. The representation in frames is symbolic and does not employ quantitative measures as means of representation. Conceptual spaces, on the other hand, *use* quantitative notions as representational tools, for example, distance measures. In this respect, Zenker [40] is correct that conceptual spaces exceed the representational power of frames.[8] This leads to the above-mentioned bias in application fields. For conceptual spaces, values with quantitative information are of higher interest than for frames. By having quantitative methods, conceptual spaces can exploit quantitative information.

The most important advantage of frames is their flexibility, particularly in the version of Petersen [27]. As explained above, a frame is a structure consisting of functional attributes and a set $V$. Elements of $V$ can occur as arguments and as values of attributes, rendering them recursive and providing frames with considerable flexibility. This is their advantage, but it also entails a shortcoming in representing the detailed internal structure of values. As a result, frames offer a powerful tool to model concepts or larger structures (events, propositions) when the attribute value structure itself is of highest importance. If the finer structure of the values is critical, frames face limitations. At this point, conceptual spaces unfold. They model the internal structure of values with as much accuracy as necessary. This advantage is apparent regarding the representation of quantitative values and in domains with fine gradations. For example, if one aims to represent a particular shade of a colour, the representation in the colour space provides more accuracy than the use of any symbolic label. Conceptual spaces also allow to capture the fuzziness of borderline cases; for example, a greenish blue or blueish green. Moreover, concepts can be captured on a sub-symbolic, pre-linguistic level. According to Gärdenfors [10, p.

---

[8] Note that frames have been enriched by further operations like comparators [23], that improve the ability to express some kinds of properties that rely on quantitative scales (e.g. that one person is older than another one). The topological and geometric notions of conceptual spaces, however, are still much more powerful.

1-2], this is actually what conceptual spaces were developed for. Finally, contrary to frames, conceptual spaces model similarity, defined as an inverse of distance. Because similarity is a central notion in understanding categorisation, it is an important advantage to be able to represent how similar values are.

To conclude, conceptual spaces provide a more fine-grained view of values than frames. In this respect, we agree with Zenker [40] that conceptual spaces are more powerful. However we reject "that frames can be recovered rather easily within the conceptual spaces model" [40, p. 82]. Though lacking expressive power concerning details, frames model complex structures between attributes and values. They are not merely conceptual spaces without geometrical notions. Table 1 provides an overview of the comparison. In brief, conceptual spaces and frames are both based on functions that assign values from a value space to objects. Frames provide a recursive structure for attributes, which we do not find in conceptual spaces. Conceptual spaces, on the other hand, associate values with a topological or geometric structure, which we do not find in frames.

|  | Frames | Conceptual Spaces |
|---|---|---|
| Cognitive representation scheme | ✓ | ✓ |
| Functional attribute value structure | ✓ | ✓ |
| Classificatory values | common | uncommon |
| Metric values | uncommon | common |
| Recursivity | ✓ | ✗ |
| Distance measures | ✗ | ✓ |

Table 1: Frames and conceptual spaces: commonalities and differences

Having two different modelling schemes with complementary advantages and disadvantages raises the question of whether they can be aggregated and what they can gain from each other. The following sections consider this question from two angles: section 5 discusses frames as means to represent the composition and decomposition of conceptual spaces, whereas section 6 investigates how conceptual spaces can be applied in probabilistically extended frames.

## 5 Frames for combing conceptual spaces

Conceptual spaces are best known for representing properties along a low number of quality dimensions, typically within a domain. However, most concepts (e.g., 'apple', 'horse', 'human') are characterised by a combination of many properties. To represent this, Gärdenfors proposes *Criterion C*:

A natural concept is represented as *a set of regions in a number of domains* together with an assignment of *salience weights* to the domains and *information about how the regions in different domains are correlated*. [10, p. 105]

Gärdenfors and Zenker [14, p. 6] acknowledge that Criterion C resembles frames but claim that it is richer since "a representation based on conceptual spaces allows one to describe the structure of concepts such that objects are more or less central representatives of a concept" [14, p. 6]. As argued in the previous section this does not imply that frames are redundant in view of conceptual spaces. In contrast, Criterion C unfolds if the theory of conceptual spaces combines with frames. Conceptual spaces can model the inner structure of (natural) properties, but frames allow to link properties to each other and to the concept.

Table 2 presents the essential idea of Criterion C, as illustrated by Gärdenfors. The first four lines (SHAPE, COLOUR, TEXTURE, and TASTE) can be directly represented in domains. The specifications for FRUIT and NUTRITION are quite complex and may need more decomposition.

| Domain | Region |
|---|---|
| Colour | Red-yellow-green |
| Shape | Roundish (cycloid) |
| Texture | Smooth |
| Taste | Regions of the sweet and sour dimensions |
| Fruit | Specifications of seed structure, flesh and peel type, etc. according to principles of pomology |
| Nutrition | Values of sugar content, vitamins, fibers, etc. |

Table 2: Criterion C: Representation of apple from [10, p. 103].

Table 2 is a simple frame structure if the domains are interpreted as attributes. For instance, COLOUR has the value 'red-yellow-green'. However, one gains further expressive possibilities by assuming frames with recursivity. This is illustrated in a frame-based representation of the first four lines of table 2 in figure 4. The values in italic letters are the ones Gärdenfors mentions in the table. They are critical for the overall appearance of an apple. For instance, the apple peel is important for COLOUR but not for TASTE.

In the frame, one can model that apples have different parts, which allows one to apply some attributes twice: COLOUR, TEXTURE, and TASTE. The values of these perceptual attributes are indeed best represented by including conceptual spaces. In this model, it is natural to assume that only the terminal nodes correspond to

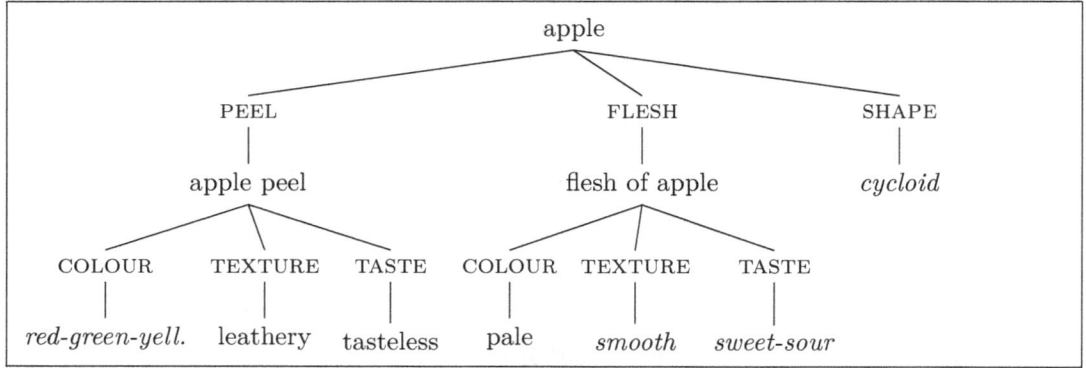

Figure 4: Criterion C as a (partial) frame of an apple.

proper conceptual spaces.[9] However, this should not be seen as a general rule for conceptual spaces in frames, as we shall see later.

The frame can easily be extended to model further information from table 2: in line 5 (FRUIT), by applying the attribute SEEDS to 'apple', yielding the value 'seeds' to which one applies STRUCTURE; in line 6 (nutrition) by applying the attribute SUGAR CONTENT to 'flesh' and so on. This simple example illustrates how frames can provide a symbolic outer structure for conceptual spaces.

Figure 4 demonstrates that frames are well-suited to represent part-whole relations. The issue of relating parts and wholes in conceptual spaces was previously addressed by Fiorini et al. [8]. They propose that a concept like 'apple' is represented as a product space, containing: 1) the holistic picture capturing the general appearance of apples; and 2) the part-whole system, consisting of a) conceptual space representation of parts (e.g., 'apple seed' in a seed space), and b) the structure space, determining the way the parts are related (e.g., that the seeds are inside the flesh).

The investigation of Fiorini et al. [8] has similarities to the frame-based proposal above, but there are also striking differences. The frame-based modelling is focused on analysing *properties* and only uses part-whole decomposing as an intermediate step to locate properties; for example, the sweet-sour taste as a property of the flesh. The proposal of Fiorini et al. [8], on the other hand, focuses on the part-whole relation itself. The most important point, however, is that frames are not only applicable to decomposing part-whole relations, but to specify all kind of attributes. One can extend the frame by including further conceptually relevant information: For example, ORIGIN of 'apple' gives 'apple tree', to which one can apply attributes

---

[9]"Proper conceptual space" means that at least one dimension exceeds a nominal level.

such as AGE in order to represent that apples grow on apple trees aged between ten and hundred years.

To sum up, the composition and decomposition of conceptual spaces in frames can go far beyond part-whole relationships. They provide a general procedure concerning how to relate conceptual spaces within a concept. As such, a frame-based approach is relevant to a long-standing problem within the theory of conceptual spaces: how are lower-dimensional conceptual spaces and more complex ones related. This question has two parts: how should one analyse complex spaces and how can conceptual spaces be combined?

A contribution to the first part of the question is the aforementioned work by Johannesson [18]. He investigates conceptual spaces with subspaces. Such conceptual spaces typically arise if several domains are intertwined in one conceptual space. The afore mentioned space of coloured dots would be an example of a combined space. That means, the dimensions in the space are *prima facie* not equally closely connected to each other. The space has integral and separable dimensions or, more precisely, integral and separable pairs of them.

The frame-based approach to conceptual spaces fits the research by Johannesson [18]. It can provide a formal structure for analysing complex spaces in terms of their subspaces, or even subsubspaces and so on, such that the ending nodes correspond to one-dimensional spaces or domains, where an approximately Euclidean metric is assumed. Spaces with city-block metrics or mixed metrics can occur higher in the paths of frames. In contrast to the example in figure 4 above, this application obviously entails that conceptual spaces appear in all nodes of the frames.

The reverse side of decomposition is the composing of conceptual spaces. Within Criterion C, Gärdenfors claims that natural concepts are a combination of properties (represented in domains). As indicated, one can formalise this idea in terms of frames. For the theory of conceptual spaces, however, the question becomes whether and how the composition yields new conceptual spaces, in which complex concepts are placed. This question has been recently addressed by Lewis and Lawry [22], who focus on the composition of prototype concepts. The foundation of their research is an understanding of concepts and properties in terms of a prototype $P$ and an uncertain threshold $\epsilon$ of tolerated deviation from the prototype. Their central aim is to formalise empirical findings on conjunct prototype nouns (e.g., 'sports that are also games') from Hampton [15]. However, within their model, they also address modifications (e.g., 'red car') and, on the most basic level, the composition of (noun) concepts from properties. Figure 5 gives the schematic illustrations of the conceptual combinations and their hierarchical structure. Lewis and Lawry [22] emphasise that their framework is quite flexible:

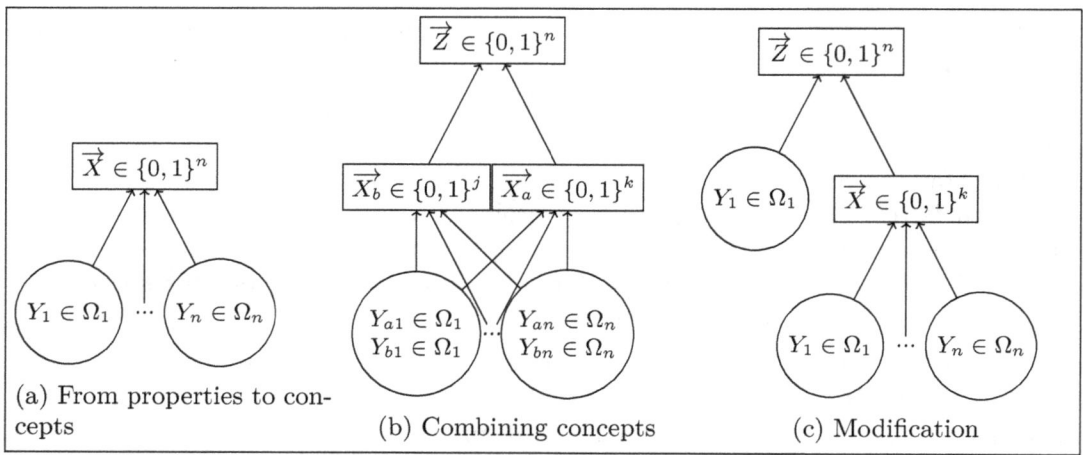

Figure 5: Schemes of conceptual combinations, taken from [22]. Properties, understood as points in conceptual spaces, can be combined to form concepts namely vectors in combination spaces (figure 5a). Higher level conceptual spaces are used to combine concepts with each other (figure 5b) or with other properties (figure 5c).

> [A]lthough when introducing this framework we have made a distinction between properties and concepts, this distinction is not really important in actually carrying out a combination. Increasingly complex concepts can be created and combined with other complex concepts or alternatively with simple properties utilising a single domain [22, p. 219].

The relaxation of the distinction between properties and concepts and the possibilities of an iterative combination brings the approach close to the recursivity in frames. The resulting hierarchical structure of the combination process in figure 5 resembles sortal frames. The schemes, however, are merely an additional illustration. They are not understood as a proper part of the conceptual representation.

A frame-based account represents the conceptual combination itself and attaches a more substantial role to it. By embedding conceptual spaces in frames, one can formally represent subspaces and combined spaces together with a symbolic representation how they are related. The complete representation of concepts consists of a graph which relates spaces and product spaces. This allows a representation of a complex concept to consist of a region in high-dimensional spaces as well as all contributing subspaces. Though such an illustration is not based purely on conceptual spaces, it suits the central aims of the theory, because a complex concept is represented in many spaces and topologically interesting phenomena can be investigated in the product spaces and the subspaces, which are hold together in the frame of the concept.

A representation of conceptual spaces and subspaces in a frame can also be applied in the discussion of how the human mind develops an understanding of domains. Gärdenfors [13] argues that most domains (i.e., their conceptual spaces) are only separated after the infant has learnt many noun concepts. The brain, he suggests, captures covariances, such as the one between flying ability, feathers and beaks, in terms of complex concepts, like 'bird', which have an overall similarity in different properties. Only later in development is the child able to single out these properties and represent them in their domains. This depiction speaks in favour of a representational tool that models relations between conceptual spaces on different levels of complexity. As argued in this section, frames can play an important role for such investigations.

# 6 Conceptual Spaces as Enrichment of Frames

Among the many concepts one could develop only some are easily learnt and efficiently used by humans. Obviously, not all possible categories one can imagine are apt for reasoning about the world. Conceptual spaces provide the means to topologically characterise what makes concepts natural. Moreover, they are related to the prototype theory of concepts. This approach, in particular the work Eleanor Rosch and her collaborators, should be given credit for drawing attention to naturalness of concepts and for shedding light on many different aspects of it, including cognitively focal points for categorisation (e.g., the typical red) [29], the role of overall similarity [31] and cognitive economy [32]. By having an inbuilt notion of centrality, distance, and similarity, conceptual spaces and prototype theory fit together well and many findings of the prototype theory can be integrated in conceptual spaces. Together with its constraint of convexity (or other topological criteria), the theory of conceptual spaces can be considered as a framework of natural categorisation.

Frames are not specialised to a particular theory of concepts. This is why they are also unrestricted with respect to the cognitive plausibility of conceptual content. In order to overcome this problem, frames need to be extended by further parameters, which allow to represent more content. This section sketches the recently researched extension by probabilities and its application to prototype concepts. The central message is that these extensions are indeed fruitful, but that they benefit from an integration of conceptual spaces and even implicitly presuppose geometric and topological structures.

**Frames and probabilities** A basic version of a probabilistically extended frame is proposed in [36]. There I suggest modelling knowledge constraints, which were already introduced by Barsalou [3], in terms of conditional probabilities. This requires that the values of attributes have a probability attached to it. Figure 6 is an example of such a frame. It represents common knowledge about birds. They have a foot structure, which is more commonly clawed than webbed. We can find birds in states of rest or movement, which are supposed to be equally common in the frame. If they move, they either walk, swim, or fly, where the latter is most likely. The constraints in the frame say that flying movement is always fast (P(fast|flying)=1) and that birds with webbed feet are more likely to move by swimming (P(swimming|webbed)=0.75). The conditional probabilities tell the agent how to adjust the likelihood if she is confronted with concepts like 'webbed-footed bird' or 'flying bird'. The details about this process and how the constraints are restricted by probability theory are discussed in [36].

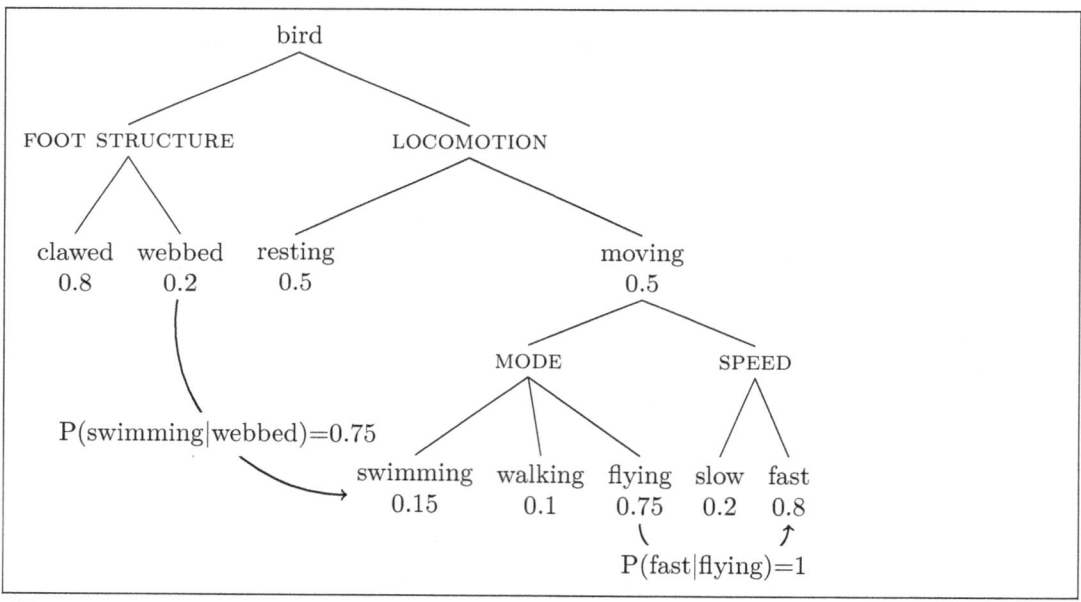

Figure 6: A partial representation of "bird" with a hierarchical structure and constraints between different levels.

The probabilistic frame allows the inclusion of more precise constraints than the original frames from [3], which only distinguish between positive and negative relevance. However, the representation is based on discrete values. This kind of frame is thus unable to represent constraints on continuous attributes, like for example our knowledge that the SIZE, WEIGHT and AGE of a person are correlated. Combining continuously valued attributes with probabilistic information requires a more general approach in terms of joint probability distributions (including probability densities).

Such a generalisation of probabilistic frames, called *stochastic frames*, has been proposed by Schuster et al. [35]. The basic idea is that the nodes are not associated to values (or types of values), but to probability distributions over possible values.[10] For example, when applying the attribute COLOUR to 'apple peel', the value is a probability distribution over the colour space. Likewise, SIZE for 'person' is a (bell-shaped) distribution over sizes.

Stochastic frames automatically incorporate constraints, namely in terms of conditional probabilities. Table 3 illustrates this for FOOT STRUCTURE and the MODE OF LOCOMOTION in the bird frame.[11] Unconditional probabilities, that is, the marginal values of the joint probability distribution, are presented in the first line and the left row. The inner part of the table displays an underlying joint probability distribution as well as conditional probabilities, that is, the constraints.

|  |  | P(fly) | P(swim) | P(walk) |
|---|---|---|---|---|
|  |  | *0.75* | *0.15* | *0.10* |
|  |  | Joint probability distribution | | |
| P(clawed) | *0.80* | 0.72 | 0.00 | 0.08 |
| P(webbed) | *0.20* | 0.03 | 0.15 | 0.02 |
|  |  | Conditional probabilities | | |
| P(...|clawed) |  | 0.90 | 0.00 | 0.10 |
| P(âĂę|webbed) |  | 0.15 | 0.75 | 0.10 |

Table 3: Probabilistic relations between MODE OF LOCOMOTION for a movement and FOOT STRUCTURE in terms of a joint probability distribution and conditional probabilities that follow from it.

An important application of stochastic frames is the modelling of vagueness in a manner previously outlined by Sutton [37]. They allow to understand the meaning of an adjective like "tall" as a shift of the probability distribution in the appropriate direction. For example, "John is tall" is not understood as giving the value 'tall' for SIZE. Rather it moves the prior probability distribution to the right. The same line of thought can be applied to colour adjectives like "green" [35]. With

---
[10]The definition was developed by Peter Sutton.
[11]Here, MODE OF LOCOMOTION refers to (the likelihood of) specific bird movements and not a bird's ability to use these, which would not be exclusive.

this application to vague adjectives, stochastic frames capture the effect of contrast classe, a phenomenon that Gärdenfors [10, p. 119–122] discusses as an application of conceptual spaces. He suggests that many adjectives operate on the area the noun concept usually occupies in the conceptual space. For example, a white wine has the lightest tones of wines have but is not white as such. Gärdenfors [10, p. 121] emphasises that this effect is difficult to address in a frame-based model. Stochastic frames overcome this limitation. However, inasmuch as stochastic frames can provide such a solution, they presuppose more geometry than is provided in frames alone.

For one thing, (joint) probability distributions over continuous variables are actually (joint) probability densities. This presupposes an underlying geometric structure of the value spaces. Moreover, the interpretation of vague adjectives as shifting values into a certain direction, whether in the colour space or in the size dimensions, anchors their meaning in this geometric structure. So far, conceptual spaces have not been explicitly included in stochastic frames, but they are implicitly present. They are not only a *possible* extension of stochastic frames but already an implicit part of them. Their role in stochastic frames, however, is to be explicated in future research.

**Prototype Frames** Prototype frames, as suggested by Schuster [34], take prototypes as weighted summary of properties.[12] They contain information about properties and their respective typicality. The according frames represent the probability of all the values an attribute can take. In this respect they are like stochastic frames but in addition they also represent attribute importance.

Attribute importance measures whether the attribute is crucial for determining category membership and is based on cue validity (cf. [30]), the conditional probability of a category $C$ given an attribute value $v_i$: $P(C|v_i)$.[13] Schuster [34] identifies attribute importance, called *diagnosticity*, by the existence of properties with a high cue validity. If $v_1, v_2, \ldots v_n$ are possible values of an attribute $A$, the diagnosticity is $diag(A|C) = max(P(C|v_1), P(C|v_2), ..., P(C|v_n))$. For example, MODE OF LOCOMOTION is important for discriminating birds from other vertebrates, because the value 'flying' distinguishes bird. In contrast, the attribute NUMBER OF EYES is unhelpful to differentiating birds from other vertebrates.

The diagnosticity formula rests on the assumption that the values $v_1$, $v_2$, ..., $v_n$ are cognitively appropriate concepts. A prominent example in [34] is the diag-

---

[12]Note that this understanding is fundamentally different from the understanding of a prototype as a central member of a category as in [22].

[13]The comparison is restricted to appropriate categories. That is, one calculates the probability of a category (e.g, 'bird') given a property (e.g., 'flying') within an appropriate superordinate category ('vertebrates') and leaves out unconnected categories ('space ships', 'planets', and so on).

nosticity of the COLOUR attribute for distinguishing fruits from vegetables. If the values were arbitrary, the formula would fail to intuitively grasp diagnosticity. By choosing inadequate properties, one could make $A$ almost arbitrarily (un)diagnostic. For example, the red colour is a distinguishing property of most fruits as opposed to vegetables, of which only few are red. This makes the colour attribute diagnostic for fruits. However, the result depends on using the right values, e.g., 'red', rather than 'red or white'. In other words, saying that a property (and its according attribute) is diagnostic makes only sense in light of an appropriate understanding of natural properties as provided in conceptual spaces.

Another connection is that attribute importance is a crucial parameter in the theory of conceptual spaces as well. Gärdenfors [10] includes it not only in Criterion C, but also as part in his weighted versions of the Minkowski distance equations. The determination of attribute importance in frames and the weighting of dimensions in conceptual spaces allows them to mutually complement each other: prototype frames offer a probabilistic definition of how to measure attribute importance while conceptual spaces offer a way to geometrically represent it.

Like conceptual spaces, prototype frames have a notion of similarity. It relates probability distributions of one attribute, usually comparing a subcategory $SC$ (e.g., 'apple') to a supercategory $C$ (e.g., 'fruit' ). If an attribute $A$ (e.g., TASTE) has $n$ possible values (e.g., 'sweet', 'sour'), then, according to [34], the similarity is:[14]

$$Sim(C, SC|A)) = \sum_{i=1}^{n} min(P(v_i|C), P(v_i|SC)).$$

For birds ($B$) and the subcategory of clawed-footed birds ($CB$) on the attribute MODE OF LOCOMOTION (short: MODE), the formula yields $Sim(B, CB|\text{MODE}) = min(0.75, 0.9) + min(0.15, 0) + min(0.1, 0.1) = 0.85$. For the subcategory of webbed-footed birds ($WB$), the formula gives $Sim(B, WB|\text{MODE}) = min(0.75, 0.15) + min(0.15, 0.75) + min(0.1, 0.1) = 0.4$. A direct comparison of 'webbed footed bird' and 'clawed footed bird' on MODE OF LOCOMOTION yields: $Sim(CB, WB|\text{MODE}) = min(0.9, 0.15) + min(0, 0.75) + min(0.1, 0.1) = 0.25$.

It is quite obvious that the introduction of similarity measurements links frames to conceptual spaces. Indeed, it seems that conceptual spaces are not only applicable *within* prototype frames for structuring the value space, but that prototype frames themselves can be placed in spaces. However, similarity in prototype frames is not a geometric notion. Any distance-based measure of (dis)similarity satisfies the three axioms on page 3. $Sim(B, C|A)$ complies with a reverse of mini-

---

[14] The formula was developed by Annika Schuster and Gerhard Schurz.

mality[15] and symmetry but the (reversed) triangle inequality [39], $Sim(B,D|A) \geq Sim(B,C|A) \times Sim(C,D|A)$ does not hold in prototype frames. In the above example, the similarity of 'webbed-footed bird' and 'clawed-footed bird' with respect to movement is only 0.25, which is less than $0.85 \times 0.4 = 0.34$. It would go beyond the scope of this paper to illustrate or even settle the dispute whether similarity *should* be a geometric notion. However, it is important to note that the similarity notion of prototype frames and of conceptual spaces follow from different traditions. Whether it is possible to modify prototype frames in a way that makes them compatible to geometric representation and the question of how this should be done are matters for future research.

**Remarks on Informativity and Naturalness** The beginning of the section pointed out that frames, in contrast to conceptual spaces, are quite unrestricted with respect to the represented content and thus barely provide the means to discuss the naturalness of concepts. One problem was that frames, as representation of categories, are quite uninformative, while informational content is relevant to deciding whether a concept is natural. According to Rosch "the task of a category system is to provide maximum information with the least cognitive effort" [29, p. 28]. Schurz [33] supports this line of argumentation through an evolutionary explanation. By the principles of variation and selective reproduction, evolution shaped an environment of similarities and correlations. Developing prototype representations promoted our survival. That means, many natural concepts are an evolutionary developed form of probabilistic reasoning.

In this regard, it is clear that probabilistic frames (including their geometric background assumptions) improve the abilities of frames to model criteria of naturalness significantly. In particular, the probabilistic extension is a prerequisite for specifying what Barsalou has dubbed constraints and what Gärdenfors calls "information about how the regions in different domains are correlated". Note that correlations are important in two ways. First, there are relations within categories, like the relation between 'webbed-feet' and 'swimming' in 'bird'. The even more crucial correlations, however, are not the ones *within* categories, but the ones that are captured *by* categories. For example, having a beak and having feathers is highly correlated within the category of animals but not within the category of birds. Almost all birds have both properties. What makes the concept 'bird' natural is that it captures these reliable correlations. From "x has a beak" one can reliably infer "x is a bird" and this makes it almost certain that x has feathers. Concepts collect highly

---

[15] $Sim(B,C|A) = 1$ if and only if the probability distributions of $C$ and $B$ on $A$ are identical and $Sim(B,C|A) < 1$ otherwise.

correlated properties. Often the correlations are very strong and instances of natural concepts can be independently identified in different attributes (e.g., many animals by SMELL, SOUND, SHAPE). Like the theoretical concepts in sciences, they seem to be "multiply operationalised". At this point, classification frames can contribute to the understanding of naturalness in concepts, because they model subcategories as result of correlations. This makes it promising to also keep in mind probabilistic versions of classification frames. In combination with conceptual spaces, they can promote our understanding of natural categories.

## 7 Conclusion

This paper had two aims: to compare conceptual spaces with frames and to outline possible integrations and fields of joint research. We argued that conceptual spaces and frames are complementary approaches. Conceptual spaces enable a fine-grained representation of quantified information and similarity judgements. Frames offer the recursive structure to relate domains and specify subcategories. Conceptual spaces and frame representation neither exclude each other nor is one framework generally more powerful. Each has advantages in respects to which the other is limited, rendering their integration into a unifying framework promising.

We discussed potential integrations in sections 5 and 6. Conceptual spaces can make use of frames to relate spaces and subspaces. This is, inter alia, relevant for representing whole-part relations or conceptual combinations. Conversely, frames, particularly probabilistic ones, require a more fine-grained representation of value spaces, which is possible in conceptual spaces.

During this investigation, it became apparent that conceptual spaces and frames already approach each other insofar as they implicitly use each other's toolboxes. Basic frame structures are found in conceptual spaces research and frame theorists implicitly assume geometric notions. Further research must not so much develop a new framework, but focus on further detailing existing overlaps in a unifying approach.

## References

[1] Janet Aisbett and Greg Gibbon. A general formulation of conceptual spaces as a meso level representation. *Artificial Intelligence*, 133(1-2):189–232, 2001.

[2] Hanne Andersen and Nancy J Nersessian. Nomic concepts, frames, and conceptual change. *Philosophy of Science*, 67:224–241, 2000.

[3] Lawrence Barsalou. Frames, concepts, and conceptual fields. In Adrienne Lehrer and Eva Feder Kittay, editors, *Frames, Fields, and Contrasts*, pages 21–74. Lawrence Erlbaum Associates, Hillsdale, 1992.

[4] Lawrence Barsalou and Christopher Hale. Components of conceptual representation. from feature lists to recursive frames. In Iven Van Mechelen, James Hampton, Ryszard Michalski, and Peter Theuns, editors, *Categories and concepts: Theoretical views and inductive data analysis*, pages 97 – 144. San Diego, CA Academic Press, San Diego, 1993.

[5] Lucas Bechberger and Kai-Uwe Kühnberger. A thorough formalization of conceptual spaces. In Gabriele Kern-Isberner, Johannes Fürnkranz, and Matthias Thimm, editors, *KI 2017: Advances in Artificial Intelligence*, pages 58–71. Springer, Cham, 2017.

[6] Joe Bolt, Bob Coecke, Fabrizio Genovese, Martha Lewis, Dan Marsden, and Robin Piedeleu. Interacting conceptual spaces I: Grammatical composition of concepts. arXiv preprint, 2017.

[7] Xiang Chen. The 'platforms' for comparing incommensurable taxonomies: A cognitive-historical analysis. *Journal for general philosophy of science*, 33(1):1–22, 2002.

[8] Sandro Rama Fiorini, Peter Gärdenfors, and Mara Abel. Representing part–whole relations in conceptual spaces. *Cognitive Processing*, 15(2):127–142, 2014.

[9] Jerry A Fodor. *Concepts: Where cognitive science went wrong*. Oxford University Press, Oxford, UK, 1998.

[10] Peter Gärdenfors. *Conceptual Spaces: The Geometry of Thought*. MIT Press, Cambridge, MA, 2000.

[11] Peter Gärdenfors. Conceptual spaces as a framework for knowledge representation. *Mind and Matter*, 2(2):9–27, 2004.

[12] Peter Gärdenfors. Convexity Is an Empirical Law in the Theory of Conceptual Spaces: Reply to Hernández-Conde. In Mauri Kaipainen, Frank Zenker, Antti Hautamäki and Peter Gärdenfors, editors, *Elaborations and Applications*, pages 77–80. Springer, Cham, 2019.

[13] Peter Gärdenfors. From sensations to concepts: a proposal for two learning processes. *Review of Philosophy and Psychology*, 10(3):441–464, 2019.

[14] Peter Gärdenfors and Frank Zenker. Editors' introduction: Conceptual spaces at work. In Peter Gärdenfors and Frank Zenker, editors, *Applications of Conceptual Spaces: The Case for Geometric Knowledge*, pages 3–13. Springer, Cham, 2015.

[15] James A. Hampton. Inheritance of attributes in natural concept conjunctions. *Memory & Cognition*, 15(1):55–71, 1987.

[16] José V Hernández-Conde. A case against convexity in conceptual spaces. *Synthese*, 194:4011–4037, 2016.

[17] Gerhard Jäger. The evolution of convex categories. *Linguistics and Philosophy*, 30(5):551–564, 2007.

[18] Mikael Johannesson. The problem of combining integral and seperable dimensions, 2001.

[19] Laura Kallmeyer and Rainer Osswald. Syntax-driven semantic frame composition in lexicalized tree adjoining grammars. *Journal of Language Modelling*, 1(2):267–330, 2014.

[20] Stephan Kornmesser. A frame-based approach for theoretical concepts. *Synthese*, 193(1):145–166, 2016.

[21] Stephan Kornmesser and Gerhard Schurz. Analyzing theories in the frame model. *Erkenntnis*, https://doi.org/10.1007/s10670-018-0078-5, 2020.

[22] Martha Lewis and Jonathan Lawry. Hierarchical conceptual spaces for concept combination. *Artificial Intelligence*, 237:204–227, 2016.

[23] Sebastian Löbner. Frame theory with first-order comparators: Modeling the lexical meaning of punctual verbs of change with frames. In Helle Hvid Hansen, Sarah E. Murray, Mehrnoosh Sadrzadeh, and Henk Zeevat, editors, *Logic, Language, and Computation*, pages 98–117, Berlin, Heidelberg, 2017. Springer.

[24] Marvin Minsky. A framework for representing knowledge. In Patrick Henry Winston and Berthold Horn, editors, *The psychology of computer vision*, volume 67, New York, 1975. McGraw-Hill.

[25] Gregory Murphy and Douglas L. Medin. The role of theories in conceptual coherence. *Psychological Review*, 92(3):289–316, 1985.

[26] David M. Perlmutter. *Deep and Surface Structure Constraints in Syntax*. New York: Holt, Rinehart and Winston, 1971.

[27] Wiebke Petersen. Representation of concepts as frames. In Jurgis Skilters, Fiornza Toccafondi, and Gerhard Stemberger, editors, *Complex Cognition and Qualitative Science. The Baltic International Yearbook of Cognition, Logic and Communication*, volume 2, pages 151–170. University of Latvia, Riga, 2007.

[28] Martin Raubal. Formalizing conceptual spaces. In Achille Varzi and Laure Vieu, editors, *Formal ontology in information systems, proceedings of the third international conference (FOIS 2004)*, volume 114, pages 153–164, Amsterdam, 2004. IOS Press.

[29] Eleanor Rosch. Natural categories. *Cognitive Psychology*, 4(3):328 – 350, 1973.

[30] Eleanor Rosch. Principles of categorization. In Eleanor Rosch and Barabar. Lloyd, editors, *Cognition and categorization*, pages 28–49. Erlbaum, Hillsdale, NJ, 1978.

[31] Eleanor Rosch and Carolyn B. Mervis. Family resemblances: Studies in the internal structure of categories. *Cognitive Psychology*, 7(4):573–605, 1975.

[32] Eleanor Rosch, Carolyn B. Mervis, Wayne D. Gray, David M. Johnson, and Penny Boyes-Braem. Basic objects in natural categories. *Cognitive Psychology*, 8(3):382–439, 1976.

[33] Gerhard Schurz. Prototypes and their composition from an evolutionary point of view. In Markus Werning, Wolfram Hinzen, and Edouard Machery, editors, *The Oxford Handbook of Compositionality*, Oxford handbooks in linguistics, pages 530–553. Oxford University Press, Oxford and New York, 2012.

[34] Annika Schuster. *Prototype frames*. PhD thesis, Heinrich Heine University Düsseldorf, 2019.

[35] Annika Schuster, Corina Strößner, Peter Sutton, and Henk Zeevat. Stochastic frames. In *Conference proceeding of "Probability and Meaning"*, ACL Anthology, 2020.

[36] Corina Strößner. Compositionality meets belief revision: a bayesian model of modification. *Review of Philosophy and Psychology*, 2020. https://doi.org/10.1007/s13164-020-00476-8

[37] Peter R. Sutton. *Towards a Probabilistic Semantics for Vague Adjectives*, pages 221–246. Springer, Cham, 2015.

[38] Ioannis Votsis and Gerhard Schurz. A frame-theoretic analysis of two rival conceptions of heat. *Studies in History and Philosophy of Science Part A*, 43(1):105–114, 2012.

[39] James M. Yearsley, Albert Barque-Duran, Elisa Scerrati, James A. Hampton, and Emmanuel M. Pothos. The triangle inequality constraint in similarity judgments. *Progress in Biophysics and Molecular Biology*, 130:26 – 32, 2017.

[40] Frank Zenker. From features via frames to spaces: Modeling scientific conceptual change without incommensurability or aprioricity. In Thomas Gamerschlag, Doris Gerland, Rainer Osswald, and Wiebke Petersen, editors, *Frames and Concept Types*. Springer, Cham, 2014.

# A Vector Simplex Model of Concepts

Douglas Summers-Stay
U.S. Army Research Laboratory
douglas.a.summers-stay.civ@mail.mil

## Abstract

Certain seemingly incompatible properties of concepts, as explored in cognitive science experiments, can be understood as compatible in a model that represents concepts as regions in high-dimensional semantic space. After briefly reviewing the main theories of concepts and some outstanding questions in the field, we lay out the structure of a vector-based memory and reasoning system. With this background, we show how many of the model accounts for many puzzling features of mental concepts.

## 1 Views of Concepts

The nature of mental representation of concepts has led to several heavily debated questions in philosophy and cognitive science. In *Concepts and Categorization* [12], Medin defines three views of concepts. The *classical* view, held from ancient Greek times up until the 1960s, is that concepts are defined by a set of necessary and sufficient properties. The *probabilistic* view is that a category is represented by a central or "average" representative of all its individual members, known as a prototype. This theory became the most prominent in the 1970s. The *exemplar* view suggests that rather than a single prototype, a category consists of a set of examples, and individual entities are compared to the entire set to determine membership. It was developed as a response to criticisms of the probabilistic theory starting in the 1980s. A fourth view, named by Adam Morton around the same time [14], is the *theory-theory*: "that concepts are organized within and around theories, that acquiring a concept involves learning such a theory, and that deploying a concept in a cognitive task involves theoretical reasoning, especially of a causal-explanatory sort" [30].

These models seem, at first glance, to contradict one another: for example, they differ in their answers to whether a concept is better modeled by many things (exemplars) or by one thing (a prototype). Yet each has points in its favor the others lack. The classical theory allows the analysis of categories by breaking them down into the intersection of properties which are in some sense simpler than the concept itself. The prototype theory allows the concept to be represented as a single thing, and explains why the most typical properties tend to be best retained in memory [20]. The exemplar theory provides a plausible explanation of how newly learned examples can influence how a concept is used to classify, and why features cluster in particular combinations [19]. The correct model of concepts, whatever it may turn out to be, must explain all these properties at once.

The papers exploring these issues usually argue for or against one of the four models of concepts mentioned above, or propose modifications or extensions to a model.[1] In our scheme, all these views can be considered as different operations performed on the same underlying structure. Concept *exemplars* are represented as vectors in a high-dimensional floating-point-valued vector space, constructed in such a way that similar exemplars have nearby vectors. The set of these exemplars, and interpolations between them, forms the *extension* of the concept. The concept *prototype* is taken to be a weighted sum of these exemplars. Individual properties can also be represented as concept prototypes: the weighted sum of all exemplars which share those properties. Sparse vector decomposition of the concept prototype can recover which individual exemplars make up a concept, reversing the averaging process. The classical theory corresponds to finding the intersection of property concepts. The exemplar theory focuses on the individual exemplar vectors. The prototype theory corresponds to analysis of the concept prototype vector as if it were itself an exemplar. Relations between concepts (such as "$A$ and $B$ imply $C$") can also be represented by vectors, by an encoding scheme that allows the system to find a tree of deductive reasoning supporting a concept by the same process of vector decomposition, supporting the theory-theory.

Exemplars, prototypes, classical intersections of properties, and theories can all be interchanged for each other as needed in the reasoning process and are simply different ways of looking at a region in semantic space and its geometric relationship with nearby or overlapping regions. Whether or not one finds these arguments for how human concepts and conceptual reasoning work compelling, there are several ideas here that might be useful to include in other systems of vector representations, including the notions of a concept simplex and sparse decomposition of semantic vectors.

---

[1] A good overview of the arguments and experiments supporting or contradicting these models is the first four chapters of [15].

## 2 Vector Representations

Semantic vectors can be created in many ways. Distributional semantic word vectors, such as word2vec [13], GloVe [18], and so forth, are widely used in computational linguistics. Vectors derived from knowledge bases, such as TransE [3] and its relatives, are a second source. Numberbatch [26] and others combine distributional and knowledge base resources to get higher quality vectors. A more recent idea is using a neural network to assign vectors to arbitrary phrases as with ELMo [10] representations. All these methods can be refined to provide representations for individual word senses [4], which is more appropriate for the construction of concepts than words themselves. Each method results in a mapping where vectors representing semantically related items have similar representations. Whenever this is the case, the vectors will also support analogical arithmetic to some extent. That is, when A is to B as C is to D, the vectors for A,B,C, and D are such that the expression $-A + B + C$ is approximately equal to D. (This follows from the fact that each of the pairs (A and B), (B and D), (C and D), and (A and C) must share a common context.) This analogical property that makes vector spaces capable of generalizing to previously unseen cases. This paper will use mainly word vectors from word2vec as exemplars simply for convenience, but for a system meant to do more than explore the possibilities of the architecture more refined methods of creating vectors should be used.

Most human mental representations of concepts must be different from representations derived from knowledge bases or large text corpora. These methods do a poor job at capturing direct sensory properties of perceptions, such as color, sound, and shape. The knowledge bases used capture few salient properties for any one exemplar compared to the richness of human concepts. The subsymbolic vector architecture described, however, can handle the way that concepts shade into one another, generalize to related cases, and appear in analogies in ways that purely symbolic representations never could.

## 3 A Concept Simplex

Few concepts can be completely enumerated. New exemplars that have never been seen before can be recognized because they are sufficiently similar to the already existing concept representation. According to some prototype theories, the vectors representing all concept exemplars that have been observed in the past are weighted and averaged together to form a prototype vector. The distance from this prototype vector is then compared to the vector for the new observation, and if these are sufficiently similar, the observation is counted as an exemplar of the concept. However, suppose a new observation comes in that looks like a cross between a flamingo and a penguin. Would this be recognized as a member of the bird concept? The vector representing this new creature could be generated by averaging the vectors for flamingo and penguin: $.5 * flamingo + .5 * penguin$. Since both flamingo and penguin are atypical birds, this new vector will be far from the prototype vector at the center of the cluster of all birds. And yet people would still be able to recognize it as a bird. Instead of a single prototype, it seems that a better representation would recognize an observation similar to any possible weighted average of exemplars from the concept.

The structure formed by taking all possible weighted combinations of a set of independent vectors is called a *simplex*. For three exemplar vectors, this would be the triangle whose corners are the exemplars. Triangle edges are weighted averages of any two of these exemplars, while the face is a weighted average of all three vectors. Weights at any vector within the triangle are known as the *barycentric coordinates* of the vector. Four exemplar vectors form a tetrahedron, and in general $n$ exemplars form an $n$-simplex. These are convex regions in semantic space, as described by Peter Gärdenfors in *The Geometry of Meaning* [9]. The individual dimensions Gärdenfors describes are typically meaningful and segregated into domains, while ours need not be. It is also similar to the spaces used by Dominic Widdows' subspace model [31]. The main difference here is that Widdows defines ($a$ OR $b$) as the space spanned by the vectors $a$ and $b$, while we restrict it to positive weights on $a$ and $b$ which sum to 1, known as the convex combination of $a$ and $b$.

The intersection of two concept simplices is the simplex formed by the exemplar vectors which belong to both concepts. When a concept simplex is incomplete, a fuzzy intersection returns exemplars which are near to both concept simplices. (That is, for exemplar $x$ and concept simplices $A$ and $B$, $max(Dist(x, A), Dist(x, B))$ is small.) When simplex $A$ is considered complete and $B$ is not, we can ask which points in $A$ are closest to $B$. For example, when querying which vehicles from a list are likely to be flying vehicles, the system measures how close each list item is to the *flying things* simplex. These methods work when all the exemplars are available. We will show in a later section, however, that intersections and unions can often be calculated even when only a prototype vector is available for each set, and decomposition into exemplars can be done afterwards.

## 3.1 Simplex Distance Experiment

We performed an experiment to determine the following: which provides a better estimate of class membership: distance to the centroid (arithmetic mean) of a set of semantic vectors belonging to that class, or distance to the simplex defined by those points? To perform this experiment, we obtained members of classes from ConceptNet. We used 1773 concepts with ten or more exemplars derived from ConceptNet, such as *IsA furniture* (*chair*, *sofa*, etc.) or *HasProperty boring* (*repetition*, *housework*, etc.). We represented the exemplars from the dataset using Mikolov's original 300-D word2vec semantic vectors. From these concepts we withheld 10% of the data as a test set. We also constructed a set of distracting semantic vectors. For each exemplar, we found a named semantic vector at the same Euclidean distance from the centroid which was not an exemplar of the class. The experiment was to decide, for each pair of exemplars, which belonged to the class and which did not. An algorithm to measure the Euclidean distance from a vector to the nearest point on a simplex was presented in [11]. In 69% of cases, the true exemplar of the class was closer to the simplex than the distractor. This indicates that distance to the simplex is a better measure of semantic nearness than distance to the mean in most cases. (See fig. 1.) Using a simplex makes the assumption that the manifold is locally Euclidean, which may not be the case. This suggests that a neural network capable of learning a nonlinear model of the structure of a concept could be a better model. However, the number of exemplars used to form a concept can be in the single digits, which is challengingly few for most learning methods. It is also possible that what we group as a class may encompass multiple distinct concepts. In that case the more appropriate representation may be a simplical complex: a collection of simplices such that the intersection of any two is a face of each of them.

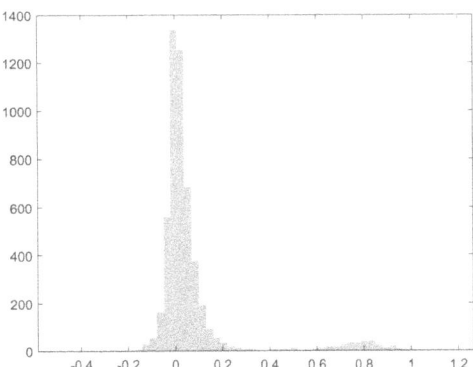

Figure 1: Histogram of the difference in distances to a concept simplex between a test vector belonging to the concept and a distractor vector that does not. Everything to the right of zero is a success (the distance is smaller for the vector belonging to the class than the distractor) while everything to the left of zero is a failure (the distance is greater for the vector belonging to the class than to the distractor.)

## 4 AND, OR, and NOT

A weighted sum of exemplars $a$ and $b$ has a meaning similar to the phrase "$a$ OR $b$" in the following sense: a person saying "I would like a sandwich or a hamburger" would likely be satisfied by something partway between a sandwich and a hamburger (a roast beef sandwich on a bun, perhaps). If exemplar weights are treated as variable, a concept simplex can thus be thought of as representing the OR-combination of all its exemplars. By contrast, "$a$ AND $b$" is better represented by the set $\{a, b\}$. To represent a concept without one of its exemplars, the exemplar can be subtracted from the sum: $a + b + c + d - a = b + c + d$. This suggests that "NOT $a$" can be represented by $-a$. Dominic Widdows has come to a similar conclusion [31], representing $a$ NOT $b$ as the portion of $a$ perpendicular to $b$: $a - \frac{a \cdot b}{|a \cdot b|} b$. (This maintains the same idea except for a multiplicative factor, which we are generally ignoring here.) These representations can be used to form the basis for a useful system for performing logical operations on concepts, as described in section 6.

## 5 Decomposition of Prototypes

Given a prototype vector for a concept, which is a weighted sum of the exemplars of the concept, is there any way to recover from this vector what the exemplars and weights are? The answer turns out to be "yes" for a surprisingly large range of cases. The problem is known as sparse vector decomposition: given a dictionary of all named vectors, the goal is to find which exemplar vectors have been summed up to make this prototype, and with what weights.(The fact that most observed vectors are not exemplars of this concept means that they will have zero weight, which is what makes this a "sparse" decomposition.) The number of vectors that can be recovered necessarily depends on the size of the dictionary, the vector dimensionality, and how related the vectors are. Surprisingly many exemplars can be recovered because most vectors only fall within a single simplex or its sub-simplices. (This holds as long as the dimensionality of the simplex is not too high compared to the dimensionality of the vectors.) We typically use a variation on non-negative LASSO to perform the decomposition. LASSO balances sparsity against exactness in finding sparse sums with a parameter, $\lambda$. It can be difficult to choose the correct lambda, so we use a screening method called DPP (Dual Polytope Projection) [29] to efficiently test over the full range of $\lambda$ values from 0 to 1, and gather all the candidate non-zero weighted vectors from this range. With a dictionary size less than or equal to the dimensionality of the vectors, it is possible to solve the resulting system of linear equations exactly. As long as the correct vectors are included among these $n$ vectors, the exact weights will be recovered.

Experiments on how complex sensory inputs such as faces and sounds are separated out in neural representations hints that the brain makes use of some kind of sparse decomposition in order to make sense of complex inputs [2]. It has been speculated that this could be achieved through competition among neural units in a winner-take-all architecture [7]. Others have found that sparsity can be achieved by appropriate thresholding [22].

Figure 2 shows how the number of vectors that can be recovered varies with vector dimensionality, holding dictionary size and relatedness constant. Up to a point, all vectors in the sum can be recovered. (This is the linearly rising part of the graph on the left.) Beyond this point, the graph still rises briefly, as there are a few errors. Once there are too many errors, however, they cascade and the number of vectors successfully recovered from the sum drops.

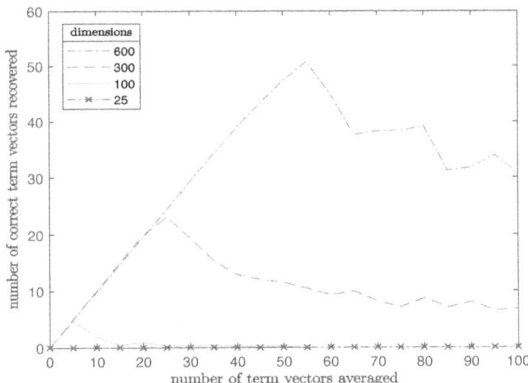

Figure 2: Number of exemplar vectors and weights that can be successfully recovered from a prototype vector of various dimensions, given a dictionary of size 100,000.

If two prototype vectors are added together, the exemplar vectors which make them up will be doubly weighted in the sum. This means that it in the case where the weights within a prototype are equal it is possible to find the union and the intersection of the exemplars based on their weights in the decomposition. Decomposition can also be performed into a dictionary of properties, returning the properties and weights by which a concept may be defined.

## 6 Decomposition for Deduction

Using vectors representing propositions $a$ and $b$, "$a$ implies $b$," is equivalent to "(not $a$) or $b$" and can be represented as the vector $-a+b$ according to the scheme described in the last section. Given that proposition $a$ is true, the truth of proposition $b$ can be derived by decomposing it into the two propositions $a$ and $(-a+b)$:

$$b = a + (-a + b)$$

Decomposition in this way can recover trees or long chains of propositional reasoning. By using a different dictionary, we can choose whether to recover exemplars or to recover a chain of reasoning. In the first case, the dictionary is a list of all named vectors. In the second case, it is a list of true proposition vectors and true logical phrases composed of them, such as "a and b and (not c) imply d".

When performing decomposition for deductive reasoning, we begin with a set of true sentences about propositions known as $KB$, for Knowledge Base. ($KB$ could also be considered as one large sentence joined by AND since it is asserting that the conjunction of each of its sentences is true.) The goal of deductive inference is to decide whether $KB \models \alpha$ for some sentence $\alpha$. The method proposed in this paper represents each of the sentences in $KB$ and the sentence $\alpha$ by one or more vectors per sentence. The vectors are constructed in such a way that the vectors corresponding to clauses which prove $\alpha$ add up to the vector representation of $\alpha$. In this way, by finding sums of vectors that add up to the vectors representing $\alpha$, we are able to find a proof that $KB \models \alpha$.

Individual propositional literals such as $A$, $B$, and $C$ are represented as basis vectors $a$, $b$, and $c$. The negation of proposition $A$, $\neg A$, is represented as $-a$. The disjunction (OR) of two or more (possibly negated) propositions, $A \vee B ... \vee C$, is represented as $a + b ... + c$. This is known as a disjunctive clause. [2] The conjunction (AND) of disjunctive clauses is not represented by a single vector but by vectors for the clauses listed separately.

Any sentence in propositional logic can be written as the conjunction of one or more disjunctive clauses. This is called conjunctive normal form, or CNF. The $KB$ as a whole is the conjunction of all of the disjunctive clauses from all the sentences that make it up. The sentence $\alpha$ is also converted into one or more vectors by converting it to CNF. We then prove $\alpha$ by finding clauses from the $KB$ that add up to the vector for each clause of $\alpha$.

---

[2] Notice that switching signs, as in logical negation, is an involution, so that $--a = a$, and that addition, like disjunction, is commutative, so that order doesn't matter. However, $\neg(a \vee b)$ cannot be encoded as $-(a+b)$ and then simplified to $-a-b$, because the distributive property does not hold. Simplification must therefore take place before encoding propositions as vectors.

Each clause vector represents its premises by negated terms and conclusions by un-negated terms. A conclusion of a previous step and a premise of the next step sum to zero, leaving only the conclusion of the final step in the sum. For example, consider proving $B$ from a $KB$ including $A$ and $A \supset B$. We represent $A$ by the vector $a$, and use the vector representation $-a+b$ for $A \supset B$. When $(-a+b)$ is added to the vector $a$, only the conclusion $b$ remains.

The contrapositive can make use of the same vector representation, since if we know that $\neg B$ is true, we can conclude $\neg A$, whose expression $-(-b)+(-a)$ also simplifies to $-a+b$. Similarly, when using the $\vee$ (OR) operator, if it is given that $\neg A$ is true, we can conclude $B$. So we subtract $-a$ and add $b$, resulting in the vector representation $-(-a)+b = a+b$. If, on the other hand, we know that $\neg B$ is true, we can conclude that $A$ is true, and the expression $-(-b)+a$ still simplifies to $a+b$.

If such a sum cannot be found, we can try to prove the negation of $\alpha$, proving that $\alpha$ is false. If neither can be proved, $\alpha$'s truth value is unknown. If both were proved, there would be a contradiction, but this cannot happen if the $KB$ has been properly encoded in CNF.

| logical operator | sentence | vector representation |
| --- | --- | --- |
|  | $A$ | $a$ |
| not | $\neg A$ | $-a$ |
| or | $A \vee B$ | $a+b$ |
| and | $A \wedge B$ | $a, b$ |
| implies | $A \supset B$ (equivalent to $\neg A \vee B$) | $-a+b$ |
| implied by | $A \subset B$ (equivalent to $A \vee \neg B$) | $a-b$ |
| equals | $A = B$ | $-a+b, a-b$ |
| not equals | $A \neq B$ | $\{a+b, -a-b\}$ |
| xor | $A \veebar B$ | $\{a+b, -a-b\}$ |
| $A$ or $B$ implies $C$ | $(A \wedge B) \supset C$ | $-a-b+c$ |
| $A$ and $B$ implies $C$ | $(A \vee B) \supset C$ | $\{-a+c, -b+c\}$ |
| $A$ implies $B$ or $C$ | $A \supset (B \wedge C)$ | $\{-a+b, -a+c\}$ |
| $A$ implies $B$ and $C$ | $A \supset (B \vee C)$ | $-a+b+c$ |

Table 1: vector representations of common expressions in propositional logic

It is possible to create more complex operations by combining these, as long as De Morgan's laws are respected. For example, $(A \wedge B) \supset C$ can be rewritten $\neg(A \wedge B) \vee C$, which simplifies, using De Morgan's laws, to $\neg A \vee \neg B \vee C$, which

715

gives the vector representation $-a + -b + c$. To find the vector representation for any sentence, we represent it in conjunctive normal form, and then replace $\neg$ with $-$, $\vee$ with $+$, and $\wedge$ with a comma, indicating multiple vectors.[3] Requiring that all sentences must be converted into CNF before being included in the $KB$ also eliminates the problems that would otherwise occur when dealing with multiple copies of the same term or its negation. For example, $A \vee B$ is represented as $a + b$, but $A \vee A$ cannot be represented as $a + a$ because it is logically equivalent to $A$ and therefore must be represented as $a$. By requiring that sentences must first be converted into CNF, such problems are dealt with in the preparation stage.

The primary benefit of using decomposition rather than traditional methods of finding chains of reasoning is flexibility in allowing concepts which have been phrased slightly differently to still connect, allowing the chain of reasoning to "go through." By embedding the entire deductive reasoning process in the vector space, the system can take advantage of associational and analogical reasoning in order to fill in gaps that would cause forwards or backwards inference to return no results. For example, in an experiment using this method reported in [27], in 548 out of 1000 cases, the top result returned was the correct one, in cases where a traditional knowledge base would have returned no answer.

Sparse decomposition is not difficult to achieve with a neural system: if a brain state is characterized as activation weights on a set of neurons, then each exemplar and prototype are represented by a particular state of these neurons, and a form of sparse decomposition has been shown to be biologically plausible [21]. Combined with recent research which "suggest[s] that empiricist, prediction-based vectorial representations of meaning are a viable candidate for the representational architecture of human semantic knowledge," [24] the architecture described here may be a candidate for a reasonably realistic model of brain representation of concepts.

# 7 Example

Here is a fully worked example from [23]. The six asserted clauses were mixed in with 94000 other clauses derived from Conceptnet to act as distractors.

> If the unicorn is mythical, then it is immortal, but if it is not mythical, then it is a mortal mammal. If the unicorn is either immortal or a mammal, then it is horned. The unicorn is magical if it is horned.

---

[3]Converting new assertions to CNF before adding them to the knowledge base is a technique commonly used in large knowledge bases such as Cyc: see http://www.cyc.com/subl-information/cyc-canonicalizer/

Interpretation:

- $Y$: it is mythical, $O$: it is mortal, $A$: it is a mammal, $H$: it is horned, $G$: it is magical
- $(Y \to \neg O) \wedge \neg Y \to (O \wedge A)$
- $(\neg O \vee A) \to H$
- $H \to G$

Everything after this point is automatic.
Conjunctive normal form:

- $(A \vee \neg O) \wedge (\neg O \vee \neg Y) \wedge (O \vee Y)$
- $(\neg A \vee H) \wedge (H \vee O)$
- $\neg H \vee G$

Vector Representation:

- $-o + a,\ -o - y,\ o + y$
- $-a + h,\ h + o$
- $-h + g$

We perform decomposition on the vector $h$, resulting in the following proof that the animal is horned:

$$\frac{1}{2}(h + o) + \frac{1}{2}(-o + a) + \frac{1}{2}(-a + h) = h$$

Decomposition of $g$ generates the following proof that the animal is magical:

$$\frac{1}{2}(h + o) + \frac{1}{2}(-o + a) + \frac{1}{2}(-a + h) + (-h + g) = g$$

There are a few things to notice about these equations. First, although we have arranged their left hand side to show the reasoning as an ordered chain, in fact the terms are returned as an unordered set, and ordering, if desired, must occur in a postprocessing step. Second, notice that some of the terms have a fractional weight on them. The system assigns these weights automatically in such a way that the final sum will equal the goal vector. For the purpose of finding out which clauses are needed for a proof, these weights can be ignored; they nevertheless have some relevance for applications using linear logic. Third, addition inside of the parentheses represents OR, while addition outside of the parentheses represents AND. This double assignment of addition works only in the specific context of deductive inference of clauses already in conjunctive normal form, because in such a context, DeMorgan's laws don't come into play, and $A$ and $\neg A$ in separate clauses cancel out as they need to.

Decomposition of $y$, an attempt to prove that the unicorn is mythical, returns the following one-step proof:
$$(o + y)$$
i.e., it is either mortal or it is mythical. Since the sum does not add up to $y$, we know that this is not a complete proof of $y$.

## 8 Vector Representation of Concepts

This vector simplex representation of concepts is the simplest possible model that has the property that any point between two known exemplars of a concept belongs to the concept. It provides a possible solution to several puzzling properties that concepts have been shown to have:

***Properties and relations can themselves be concepts. [5]***

The extension of a property may be represented by the region spanned by all exemplars which share that property. Since "$a$ implies $b$" can be represented by the vector $-a + b$ in this scheme, a concept formed of the convex space spanned by a set of such vectors is itself a concept.

***Concepts are organized in such a way that a concept similar to any two other concepts can be found.***

The nature of high-dimensional vector spaces is such that all vectors are close together in the sense that $(a + b)/2$ is closer to both $a$ and $b$ than to any other observed vectors in the dictionary, for any reasonable dictionary size.

For instance, the word *cartoon* is closest to the word *cartoons*: their dot product is .80. A distant word is *cardiology*: the dot product with *cartoon* is .02. *Cardiology* is professional, serious, and correlated with aging, while *cartoons* are entertainment, silly, and related to children, so they would seem to have little to do with each other, and be far apart in the semantic vector space. And yet, the closest results to the midpoint of the two terms are *cartoon* and *cardiology*. In other words, even for distant words there is a point which is closer to those two terms than to any other terms in the dictionary. Other terms close to this vector include *pediatric cardiology* and *editorial cartoon*, which exchange certain features of the term with the other term (youth, in the first case, and age and seriousness in the second). If we took such distant terms in a two-dimensional semantic vector space, their midpoint would be somewhere in the middle of the map and be surrounded by many completely unrelated words. In a high-dimensional semantic vector space, however, any point along the line connecting any two distant terms is only nearby terms which are similar to one or both of the terms. This means that *any* two concepts can be combined to form a new concept with properties between the parent concepts.

***Exemplars can themselves be broken down into exemplars (bird exemplars might include robin, which itself might include individual robins or robin sightings).***

The possibility for prototype vector decomposition into exemplars lets us form multi-level hierarchies where the concept's exemplars are prototypes formed from weighted averages of exemplars at the next layer down.

***A concept prototype changes based on context or perspective.*** [1]

Whenever a prototype is needed which is different in some way, it can be recreated by choosing a different weight set and re-averaging the exemplars. In these experiments [1], participants were asked to consider a category from another culture's point of view. Adding the vector for the culture to the prototype, and recalculating the weights on exemplars needed to decompose this modified prototype vector, can give similar results.

Taking weighted averages of vectors is a common process in computational linguistics. However, this has generally been treated as a kind of summarization of meaning that inevitably loses track of the vectors that make up the sum – a lossy compression. [28] shows that vectors that make up such a sum are recoverable when they are few compared with the vector's dimensionality, and depending on how closely related they are.

***Concepts can be defined by their relation to other concepts, which themselves seem to be defined by their relation to the first concept.*** [8]

When treating concepts as separate entities, facts about the relations between concepts are difficult to place: "In general, it may be extremely difficult, if not impossible, to identify where the knowledge for a particular category in long-term memory begins and ends. To the extent this is true, it is hard to imagine how there could be invariant representations for categories stored in long-term memory." [16]

Definition circularity is problematic if one is trying to construct a world model by starting from only a few "primitive concepts." In contrast, if many concepts are being added to the mental space by a process of simply considering usage contexts, then sorting into regions where the same properties hold will happen automatically, and if the concepts are already in place, then the concept *bachelor* will be observed, as desired, near the intersection of the regions *unmarried* and *man*. Indeed, without such an ability, how could the first dictionaries have possibly been written? Concepts, their properties, the relations between them, and the means of reasoning over them must all be encoded in the same structures for our behavior in regards to concepts to be explained.

***Despite goldfish not being a highly prototypical fish or a highly prototypical pet, it is a highly prototypical pet fish. [17]***

Take the intersection of the *fish* concept simplex with the *pet* concept simplex. The intersection is itself a concept simplex, with exemplars which are all both pets and fish. Taking an evenly weighted average of these exemplars, gives a prototype vector for *pet fish*. *Goldfish* is an exemplar which is near to that prototype. The idea that composition of vector representations of concepts can solve the pet fish problem was explored in [6].

***There can be prototypes, exemplars, and properties defining concepts, and yet each time the same subject is asked what they are, slightly different answers are generated. [1]***

We hypothesize that the answers people give to questions about prototypes, exemplars, and properties are not the precise vectors making up the concept in our brain. Instead, they are generated anew from the mental representations as needed. Given exemplars, the brain can look for or generate prototypes and properties; given prototypes, it can look for exemplars or properties; and given properties, it can come up with exemplars or prototypes.

***Concepts can be used for inference and conceptual combination as well as classification. [25]***

A key questions that cognitive science has had little to say about yet is how it is possible for concepts to represent both concrete ideas and rules for how those ideas are related. Our model provides a method by which a vector can represent a rule for combining other vectors, allowing the same structure (a vector) to serve as both an object to be reasoned about and the rules that constrain that reasoning. Finding chains of reasoning by vector decomposition shows how concepts can form linked paths from one to another.

***Learning new information about an exemplar can update all relevant concepts at once. [25]***

If prototype vectors as being recomputed on the fly from exemplars as needed, then updating one exemplar's position as we learn more about it will modify to a greater or lesser extent prototypes calculated from sums involving it. In our model, these prototypes are not fixed entities but the result of a calculation, which means that changing one of the terms of the calculation will directly affect all the concepts in which it participates.

## 9 Conclusion

Treating a concept as a simplex of exemplar vectors in semantic space, along with functions to find a prototype vector for the concept, to recover exemplars from the prototype, and to find chains of reasoning between vectors in this space, creates a representation that is capable of reproducing at least some of the seemingly puzzling behavior of human mental concepts. We have introduced the notion of distance to the simplex as a principled way of estimating whether a new exemplar belongs to a class concept, and have shown how our previous experiments in vector-based reasoning lend theoretical and perhaps also empirical pllausibility to such a representation.

## References

[1] Lawrence W Barsalou. The instability of graded structure: Implications for the nature of concepts. *Concepts and conceptual development: Ecological and intellectual factors in categorization*, 10139, 1987.

[2] Michael Beyeler, Emily Rounds, Kristofor Carlson, Nikil Dutt, and Jeffrey L Krichmar. Sparse coding and dimensionality reduction in cortex. *BioRxiv*, page 149880, 2017.

[3] Antoine Bordes, Nicolas Usunier, Alberto Garcia-Duran, Jason Weston, and Oksana Yakhnenko. Translating embeddings for modeling multi-relational data. In *Advances in neural information processing systems*, pages 2787–2795, 2013.

[4] José Camacho-Collados and Mohammad Taher Pilehvar. From word to sense embeddings: A survey on vector representations of meaning. *CoRR*, abs/1805.04032, 2018.

[5] Nino B. Cocchiarella. Logical investigations of predication theory and the problem of universals. *Journal of Symbolic Logic*, 53(3):991–993, 1988.

[6] Bob Coecke and Martha Lewis. A compositional explanation of the 'pet fish' phenomenon. In *International Symposium on Quantum Interaction*, pages 179–192. Springer, 2015.

[7] Robert Coultrip, Richard Granger, and Gary Lynch. A cortical model of winner-take-all competition via lateral inhibition. *Neural networks*, 5(1):47–54, 1992.

[8] Russell Dancy. Speusippus. In Edward N. Zalta, editor, *The Stanford Encyclopedia of Philosophy*. Metaphysics Research Lab, Stanford University, winter 2016 edition, 2016.

[9] Peter Gärdenfors. *The geometry of meaning: Semantics based on conceptual spaces.* MIT Press, 2014.

[10] Matt Gardner, Joel Grus, Mark Neumann, Oyvind Tafjord, Pradeep Dasigi, Nelson Liu, Matthew Peters, Michael Schmitz, and Luke Zettlemoyer. Allennlp: A deep semantic natural language processing platform. *arXiv preprint arXiv:1803.07640*, 2018.

[11] Oleg Golubitsky, Vadim Mazalov, and Stephen M Watt. An algorithm to compute the distance from a point to a simplex. *ACM Commun. Comput. Algebra*, 46:57–57, 2012.

[12] Douglas L Medin and John D Coley. Concepts and categorization. *Perception and cognition at centuryâĂŹs end: Handbook of perception and cognition*, pages 403–439, 1998.

[13] Tomas Mikolov, Ilya Sutskever, Kai Chen, Greg S Corrado, and Jeff Dean. Distributed representations of words and phrases and their compositionality. In *Advances in neural information processing systems*, pages 3111–3119, 2013.

[14] Adam Morton. *Frames of Mind: Constraints on the Common-Sense Conception of the Mental*. Oxford University Press, 1983.

[15] Gregory Murphy. *The big book of concepts*. MIT press, 2004.

[16] U. Neisser. *Concepts and Conceptual Development: Ecological and Intellectual Factors in Categorization*. Emory Symposia in Cognition. Cambridge University Press, 1989.

[17] Daniel N Osherson and Edward E Smith. On the adequacy of prototype theory as a theory of concepts. *Cognition*, 9(1):35–58, 1981.

[18] Jeffrey Pennington, Richard Socher, and Christopher Manning. Glove: Global vectors for word representation. In *Proceedings of the 2014 conference on empirical methods in natural language processing (EMNLP)*, pages 1532–1543, 2014.

[19] Gualtiero Piccinini and Sam Scott. Splitting concepts. *Philosophy of Science*, 73(4):390–409, 2006.

[20] Eleanor Rosch, Carolyn B Mervis, Wayne D Gray, David M Johnson, and Penny Boyes-Braem. Basic objects in natural categories. *Cognitive psychology*, 8(3):382–439, 1976.

[21] Christopher Rozell, Don Johnson, Richard Baraniuk, and Bruno Olshausen. Locally competitive algorithms for sparse approximation. In *Image Processing, 2007.*, volume 4, pages IV–169, 2007.

[22] Christopher J Rozell, Don H Johnson, Richard G Baraniuk, and Bruno A Olshausen. Sparse coding via thresholding and local competition in neural circuits. *Neural computation*, 20(10):2526–2563, 2008.

[23] Stuart J Russell and Peter Norvig. *Artificial intelligence: a modern approach*. Malaysia; Pearson Education Limited, 2016.

[24] Jona Sassenhagen and Christian J Fiebach. Traces of meaning itself: Encoding distributional word vectors in brain activity. *bioRxiv*, page 603837, 2019.

[25] Karen O Solomon, Douglas L Medin, and Elizabeth Lynch. Concepts do more than categorize. *Trends in cognitive sciences*, 3(3):99–105, 1999.

[26] Robert Speer, Joshua Chin, and Catherine Havasi. Conceptnet 5.5: An open multilingual graph of general knowledge. In *Thirty-First AAAI Conference on Artificial Intelligence*, 2017.

[27] Douglas Summers-Stay. Propositional deductive inference by semantic vectors. In *Proceedings of SAI Intelligent Systems Conference*, pages 810–820. Springer, 2019.

[28] Douglas Summers-Stay, Peter Sutor, and Dandan Li. Representing sets as summed semantic vectors. *Biologically inspired cognitive architectures*, 25:113–118, 2018.

[29] Jie Wang, Jiayu Zhou, Peter Wonka, and Jieping Ye. Lasso screening rules via dual polytope projection. In *Advances in Neural Information Processing Systems*, pages

1070–1078, 2013.

[30] Daniel A. Weiskopf. The theory-theory of concepts. https://www.iep.utm.edu/th-th-co/.

[31] Dominic Widdows and Trevor Cohen. Reasoning with vectors: A continuous model for fast robust inference. *Logic Journal of the IGPL*, 23(2):141–173, 2014.

# Concept Functionals

Vincent Wang
*University of Oxford, Department of Computer Science, Quantum Group*
`vincent.wang@cs.ox.ac.uk`

### Abstract

$[-1, 1]$-valued functionals allow the semantic modelling of concepts as fuzzy and nonclassical predicates over a rich collection of domains, whilst maintaining compatibility with logical operations such as negation. We integrate this semantics with the Categorical Compositional Meaning programme, allowing us to compose and compute with concepts: in particular, we demonstrate how we may model spatial inference from vague and negated information obtained from fragments of natural language.

## 1 Introduction

We focus on two of the guiding questions in Cognitive Science. How should we represent concepts? How can they be composed to form new concepts? Gärdenfors' *Conceptual spaces theory* [12] is a framework to address the former, where concepts are modelled as convex subsets of abstract spaces. One appreciates the strength of this proposal and the naturality of convexity by the following example: if two pigments are both Red, it ought to follow that any mixture of these two pigments is also Red. To address the latter question, following the Categorical Compositional Meaning programme [16, 10, 6] – which broadly aims to elucidate the interacting compositional structure of syntax and semantics of natural language [22] – Bolt *et al.* [4] imbue Gärdenfors' conceptual spaces with structure both necessary to model concept composition, and sufficient to model how natural language directs this composition: allowing the computation of the meaning of phrases from the meanings of consituents.

We address the following gaps in the state-of-the-art. We will be able to model non-convex concepts (Figure 1). We will be able to model concepts fuzzily, as in [23](Figure 2), granting concepts truth values on a spectrum between true and false. Finally, we will not only be able to model the negation of concepts, but do so in psycho-linguistically faithful manners, capturing the following:

1. *Negation might be non-involutive:* the phrase "I am happy" may carry a different meaning than "I am not unhappy", even if we take unhappy and not happy to have the same meaning.

2. *Negation might not obey the Law of Excluded Middle:* if I only have a partial understanding of a concept, such as that of Good Music, even if I am unable to classify a new song as Good Music, does not mean that I immediately classify it as Not Good Music; some songs I might be unable to classify as either!

3. *The negation of a concept might not be a concept:* while Not Red is not a colour, Blue, Green and Yellow are colours that are certainly Not Red, so Not Red behaves more as a collection of other concepts in the abstract concept-domain of colour; yet, we can make sense of "The coffee is bitter and (not sweet).", where bitter and sweet both belong to an abstract domain of taste, so even if negated concepts are not concepts *per se*, they must still interact meaningfully with non-negated concepts.

First we will introduce the concepts underpinning the Categorical Compositional Meaning programme we work in, which will illustrate that the main challenge we face is finding a suitable compact closed category for our functional-spaces to act as a semantic category. Then we will build up the formalism behind these functional-spaces and explore how we may define concepts and negation with them. Finally we will place concepts in the category **ConcFun**, and provide an example of meaning computation.

## 2 Categorical Compositional Meaning

The general outline of the categorical compositional approach is to establish a compositional structure – interpreted categorically as the **Grammar Category** – in tandem with meaning/concept spaces – again organised categorically as the **Semantics Category** – both chosen such that functors from the grammar to semantics categories map grammatical type-reductions in the grammar to algorithms for composing meanings in the semantics. We will use Lambek pregroups [16, 17] as our grammatical compositional structure.

**Definition 2.1** (Pregroups). *A **pregroup** is a tuple $(A, \cdot, 1, -^L, -^R, \leq)$, where*

- $(A, \cdot, 1, \leq)$ *is a partially ordered monoid*
- $-^L, -^R$ *are functions $A \to A$ such that for all $a \in A$:*

$$a \cdot a^R \leq 1 \leq a^R \cdot a \qquad a^L \cdot a \leq 1 \leq a \cdot a^L$$

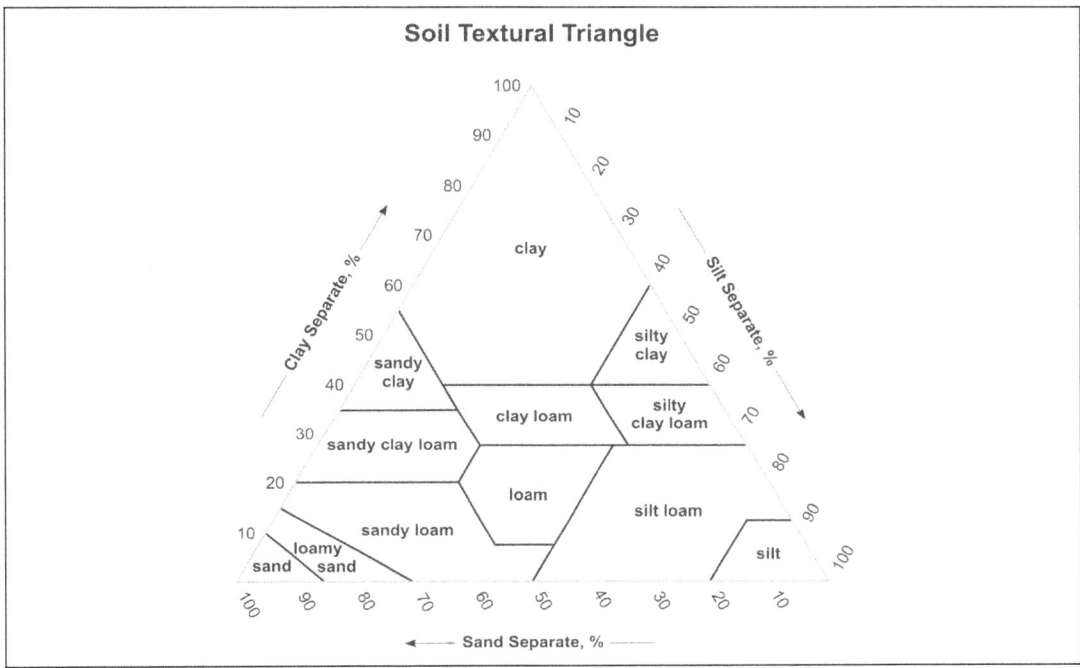

Figure 1: Soil Texture can be considered as simplicial convex domain parameterised by {Clay,Silt,Sand}: Sandy Loam and Silt Loam are nonconvex concepts; it is not the case that any mixture of Sandy Loam remains Sandy Loam: it is possible to produce Loam.

Figure 2: The television test-pattern illusion: covering the divide between any two adjacent gray patches (try with your finger) makes them appear to be the same colour. Hence Black, Grey, and White are best modelled as fuzzy concepts, where particular shades may enjoy partial membership in multiple colour-concepts.

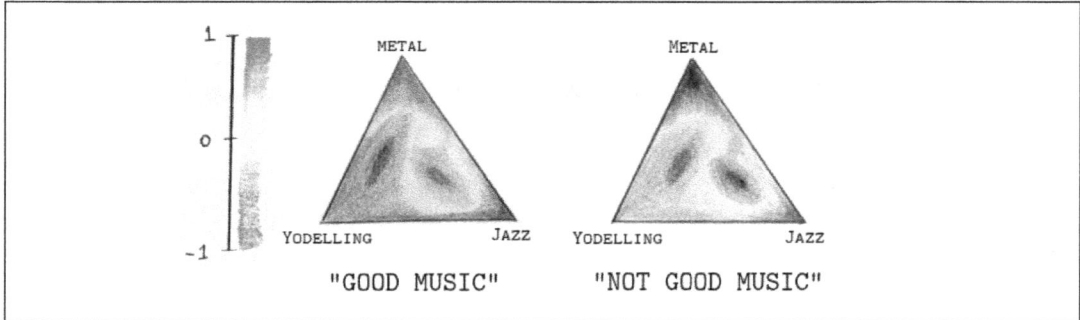

Figure 3: As a sneak peek, here is what our concepts and negations can look like.

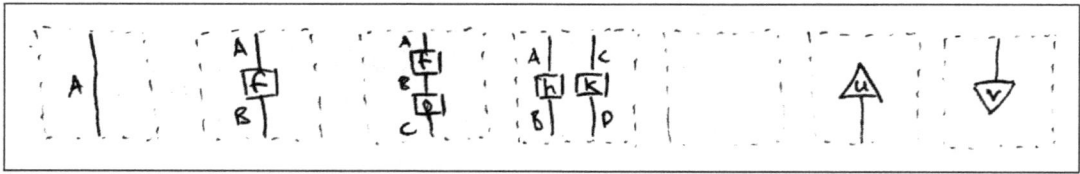

Figure 4: From left to right: an object $A$; a morphism $A \xrightarrow{f} B$; the composite $A \xrightarrow{f} B \xrightarrow{g} C$; the tensor of $A \xrightarrow{h} B$ and $C \xrightarrow{k} D$; $I$ is the empty diagram; a **state** morphism $u : I \to A$; an **effect** morphism $v : A \to I$.

Lambek pregroups can be viewed as thin rigid categories[17, 8], of which compact closed monoidal categories are a special case – there is no distinction made between the left and right adjoints $-^L, -^R$. In this paper, we will make use of compact closed categories as modelling vehicles.

**Monoidal categories** are tuples $(\mathcal{C}, \otimes, I, \alpha, \lambda, \rho)$ of a category, a **tensor**, and the **associator**, **left-** and **right-unitor** natural transformations, and that monoidal categories admit a sound and complete graphical calculus [19].

A monoidal category is **symmetric** when it admits a 'twist' natural transformation $X \otimes Y \to Y \otimes X$ that is self inverse, depicted graphically as a crossing pair of wires.

A monoidal category is **rigid** if for each object $X$ there are objects $X^L$ (the **left dual**), $X^R$ (the **right dual**), and natural transformations:

$$\eta_X^L : I \to X \otimes X^L \qquad \eta_X^R : I \to X^R \otimes X$$
$$\epsilon_X^L : X \otimes X^L \to I \qquad \epsilon_X^R : X^R \otimes X \to I$$

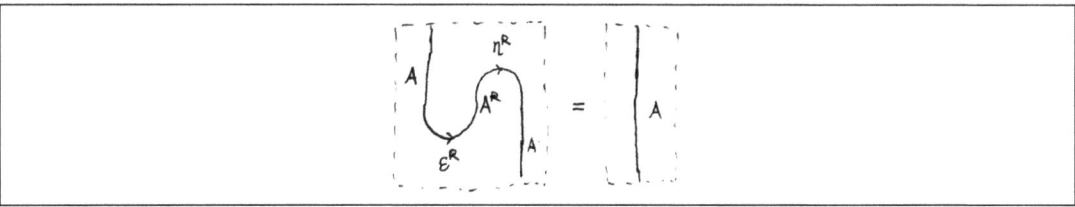

Figure 5: The depiction of the yanking equation $(\epsilon_A^R \otimes \text{id}_A) \circ (\text{id}_A \otimes \eta_A^R) = \text{id}_A$ in the graphical calulus. The $\epsilon, \eta$ morphisms are depicted as **cups** and **caps** respectively.

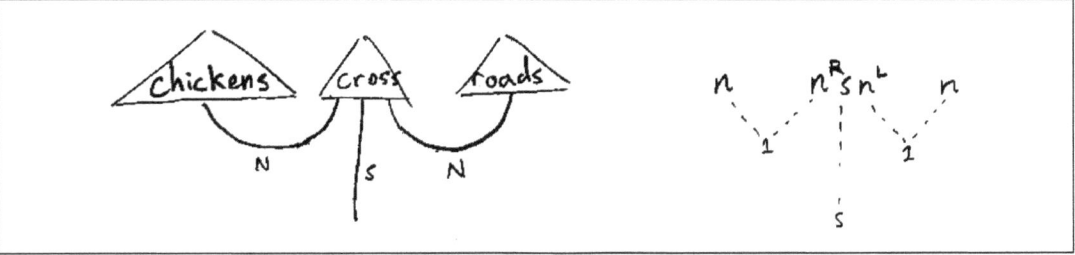

Figure 6: The sentence chickens cross roads has type $n(n^R s n^L) n$, which reduces as follows on the right: $n(n^R s n^L) n = (n n^R) s (n^L n) \leq s(n^L n) \leq s$, corresponding to the cups in the left diagram. When we can interpret the diagram on the left in a compact closed category, we can consider the meaning of the sentence to be the compound state achieved by wiring the word-states together.

Which satisfy the **yanking equations**:

$$(\text{id}_X \otimes \epsilon_X^L) \circ (\eta_X^L \otimes \text{id}_X) = \text{id}_X \qquad (\epsilon_X^R \otimes \text{id}_X) \circ (\text{id}_X \otimes \eta_X^R) = \text{id}_X$$
$$(\epsilon_X^L \otimes \text{id}_{X^L}) \circ (\text{id}_{X^L} \otimes \eta_X^L) = \text{id}_{X^L} \qquad (\text{id}_{X^R} \otimes \epsilon_X^R) \circ (\eta_{X^R}^R \otimes \text{id}_{X^R}) = \text{id}_{X^R}$$

As an example of the Categorical Compositional approach in action, consider a pregroup grammar generated by the linguistic types $\{n, s\}$, for nouns and sentences respectively; so our grammar category is $\mathbf{Preg}_{\{n,s\}}$. Suppose we have the words {chickens,cross,roads}, where chickens and roads are assigned the type $n$, and cross is assigned the type $n^R \cdot s \cdot n^L$. In Figure 6, we demonstrate how a type reduction in the pregroup corresponds neatly to the graphical calculus of monoidal categories, allowing us to evaluate the meanings of sentences.

## 3 Towards spaces of functionals

A Functional on a space can be thought of as a "truth-value field"; a truth value is assigned to each point of the underlying space. Functionals admit linear algebra, in

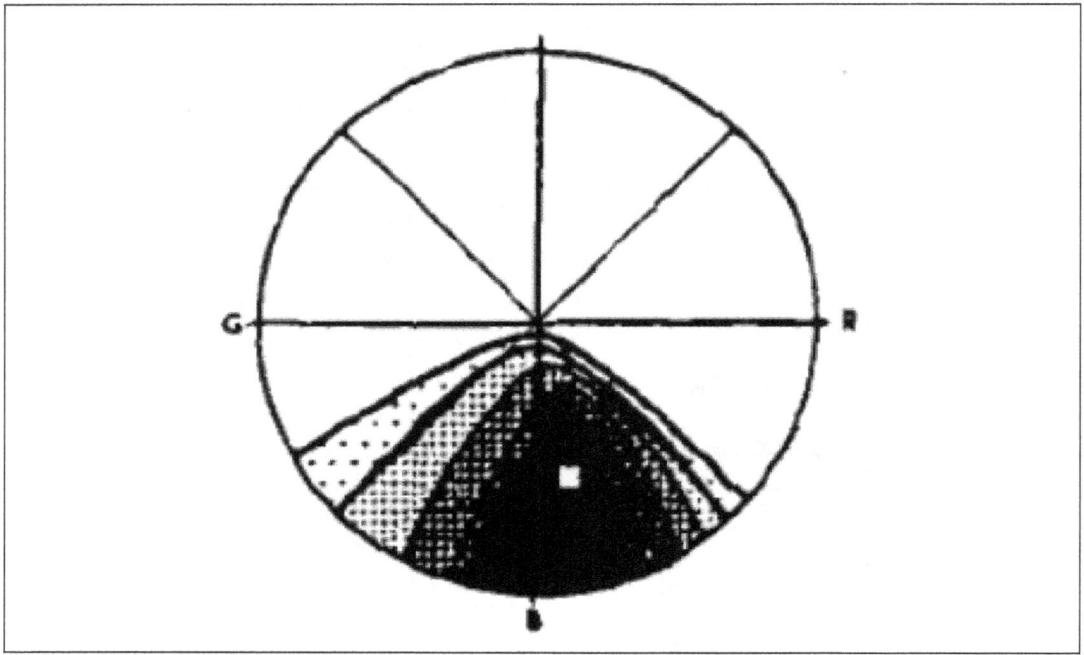

that two functionals over the same base space can be summed, and functionals can be multiplied by scalars, by summing and scaling truth-values pointwise whenever the truth-values admit those operations. For the reader who wishes to skip technicality: if we accept that concepts live in ambient geometric spaces, where points of the space are concrete instances of phenomena, functionals over those ambient spaces allow one to expressively specify, for each concrete phenomenon, to what degree that phenomenon is an instance of a concept.

**Example 3.1.** *As a first example, consider this graphic from the Scandinavian colour-naming survey by [20]. Depicted is a colour wheel, overlaid with shaded regions, where darker shades correspond to higher mean degree of agreement (on a scale from 0 'not at all' to 7 'extremely well') of how well colour samples from those regions coincided with "what [participants] mean by blue". We may model the concept 'blue' according to this data by a $[0, 7]$-valued functional on the disk.*

Functionals will be the objects of our eventual semantic category: (Hilbert) spaces of $[-1, 1]$-valued functionals over unit-measure spaces. Expressing functionals as elements of Hilbert spaces grants access to the inner product bilinear form, which gives us a way to quantify the distance between concepts. The 'unit-measure spaces' requirement is technical: in order to define the inner products in these Hilbert spaces, we must first ensure that the functionals we define admit a notion of integration, for which we require measure-theoretic notions, which we recount briefly [5].

In detail, the motivation for having integration is to grant a systematic method to map functionals onto $\mathbb{R}$, which we may then use to define notions of similarity via inner product and how far a given functional is from a classical predicate that assigns either True or False to every point in its domain. The following section recounts the mathematical boilerplate necessary, and may be skimmed or skipped without conceptual loss.

A **measure space** is a tuple $(X, \mathcal{X}, \mu)$ of a set $X$, a **measure** $\mathcal{X}$ which is a $\sigma$-**algebra** on $X$ – a collection of subsets of $X$ that contains $X$ and is closed under complement and countable unions –, and a **measure function** $\mu$ which is a set-function $\mathcal{X} \to \mathbb{R}$ that is non-negative, maps $\emptyset$ to 0, and for all countable collections $\{E_i : i \in \mathcal{I}\}$ of pairwise disjoint sets in $\mathcal{X}$, $\mu(\bigcup_{i \in \mathcal{I}} E_i) = \sum_{i \in \mathcal{I}} \mu(E_i)$. When $X$ is a topological space, and the $\sigma$-algebra $\mathcal{X}$ contains the topology of $X$, we call the measure a **Borel measure**. A **measurable function** between measure spaces $(X, \mathcal{X}, \mu_X) \xrightarrow{f} (Y, \mathcal{Y}, \mu_Y)$ is such that the preimages of measurable sets are measurable: for all $E \in \mathcal{Y}$, $f^{-1}(E) \in \mathcal{X}$. When topological spaces are equipped with a Borel measure, we can define the integrals of arbitrary measurable functions from them to $\mathbb{R}$, which we write (with respect to the symbol names above), as $\int_X f \, d\mu$. As is standard, we assume that the reals $\mathbb{R}$ carry the Lebesgue measure, where $\mu_{\mathbb{R}}([k, k+1]) = 1$ for any unit interval. The Lebesgue measure can be extended to any cartesian space $\mathbb{R}^n$, assigning measure 1 to any unit hypercube.

**Example 3.2** (All convex algebras on finite sets $X$ are fair game). *Convex Algebras were employed for conceptual modelling in [3]. In brief, convex algebras generate convex spaces from (among other structures) finite sets, by considering formal convex sums of elements. Formally, a* **convex algebra** *on a set $X$ is a tuple $(D(X), \alpha)$, where $D(X) := \{\sum_{x \in X} p_x \ket{x} : \sum_{x \in X} p_x = 1, \forall x \in X (0 \leq p_x \leq 1)\}$ is the set of formal* **convex sums** *of the set $X$. The ket-notation is used to stress the formal nature of the sums. The* **mixing operation** $\alpha : D(X) \to X$ *satisfies:*

$$\alpha(\ket{x}) = x \qquad \alpha\big(\sum_{i,j} p_i q_{i,j} \ket{x_{i,j}}\big) = \alpha\bigg(\sum_i p_i \ket{(\sum_j q_{i,j} \ket{a}_{i,j})}\bigg)$$

*We can borrow the topology and measure of $\mathbb{R}^{|X|}$. When the underlying $X$ is finite, we can define the function:*

$$y \mapsto \{ \sqrt[|X|]{|X|!} \cdot \vec{p_x} \in \mathbb{R}^{|X|} : \sum_{x \in X} p_x \ket{x} = y\}$$

that maps elements $y \in D(X)$ to the coefficients of the convex mixtures that yield $y$, viewed as convex subsets of unit-measure simplices in $\mathbb{R}^{|X|}$, so obtaining a measure and a quotient topology for $D(X)$.

We will restrict our attention to normalised Borel-measure spaces $(X, \mathcal{X}, \mu)$ with unit total measure: $\mu(X) = 1$. We wish to consider an appropriate (Hilbert) space of all such functionals for a given Borel unit-measure space $\Delta$. Recall that a Hilbert space is a vector space $V$ equipped with an inner product $\langle \_, \_ \rangle : V \times V \to \mathbb{C}$ which is sesquilinear and conjugate-symmetric [18].

**Definition 3.1** ($L^2(\Delta)$-space). *Where $\Delta$ is a Borel unit-measure space, $L^2(\Delta)$ is the Hilbert space of all functionals $f : \Delta \to \mathbb{R}$ that are square integrable: $\int_X f^2 \, \mathrm{d}\mu < \infty$. The inner product $\langle \_, \_ \rangle$ is*

$$\langle f, g \rangle = \int_X fg \, \mathrm{d}\mu$$

*$L^2(\Delta)$ is a vector space over $\mathbb{R}$, and the inner product satisfies sesquilinearity and conjugate-symmetry.*

## 4 Functionals as Concepts

We wish to normalise our functionals to take values in $[-1, 1]$, such that we at once obtain fuzzy representations of positive and negative extensions of concepts. In general, functionals may take on values outside of this range, and may be poorly behaved, tending to diverge at points to arbitrarily large values. For our purposes, we may consider functionals to be bounded.

Given a functional taking bounded values, we may normalise to obtain a functional taking values in the range $[-1, 1]$. For instance, we may take some monotone homeomorphism, denoted $\kappa$, between $\mathbb{R}$ and the interval $(-1, 1)$ such that $\kappa(0) = 0$ (such as families of sigmoids often used in Machine Learning), we may equivalently treat $L^2(\Delta)$ to be the space of square-integrable functionals on $\Delta$ taking values in $(-1, 1)$. We will instead instead identify functionals $f$ with the first $[-1, 1]$-bounded functional obtainable from $f$ by multiplication with a suitable scalar.

**Definition 4.1** (Concepts). ***Concepts*** *are functionals that take values in the interval $[-1, 1]$. The positive-valued domain of a concept is its (possibly fuzzy)* **positive extension**, *the negative-valued domain of a concept is its (possibly fuzzy)* **negative extension**[1], *and the 0-valued domain of a concept is its* **penumbra**. *Given an*

---

[1] A logic-folkloric term for tuples in a model for which a predicate evaluates to false

*arbitrary functional $f$, we define its concept $\hat{f}$ to be:*

$$\hat{f} := \frac{f}{\max\{\sup_{x \in X} |f(x)|, 1\}}$$

$|\frac{f}{\sup_{x \in X} |f(x)|}|$ takes values bounded in the interval $[-1, 1]$, so $\|\hat{f}\| \leq 1$ and hence

$$\int_X \hat{f} \, d\mu < 1$$

Now we explore the kinds of concepts we can express. The operative intuition here is that we treat a concept – or predicate – as a test on points of the domain space, which returns a value in $[-1, 1]$ corresponding to the degree to which the tested instance satisfies the predicate.

To take a simple example, the predicate 'tall', applied to people, is fuzzy, increasing in truth value (from 0 to 1, in Zadeh's fuzzy setting) as height increases. As is, for instance, 'short', which we would treat as an antonym to 'tall', decreasing (from 1 to 0) as height increases. Modelling 'tall' as a concept functional over the real line, we may take the truth value at each point to be that of 'tall' minus 'short': negative values correspond to heights that are 'short', positive to 'tall', and 0 values to heights at which the truth values for 'tall' and 'short' coincide.

**Remark 1.** *Concepts are closed under taking convex combinations, so it is already possible here to implement them in* **ConvexRel** *[4]. We will push slightly further in order to capture negation.*

## 4.1 Modelling Classical, Fuzzy, and Nonclassical Concepts

**Definition 4.2** ($\epsilon$-crisp concepts). *$f$ is an $\epsilon$-**crisp concept** when*

$$\|\hat{f}\| \geq 1 - \epsilon$$

When $\epsilon = 0$, we force $\int_X \hat{f} \, d\mu = 1$, as $\|\hat{f}\| = 1 = \int_X \hat{f}^2 \, d\mu$, and since $\hat{f}$ is bounded in $[-1, 1]$ and $X$ has unit measure, we must have that $\hat{f}$ takes values -1 or 1 almost everywhere: that is, except on sets of measure 0. In other words, these are **classical concepts** with non-fuzzy truth values, whose positive and negative and negative extensions collectively cover the domain. In other words, the Law of Excluded Middle is obeyed.

When $\epsilon > 0$, we admit penumbras and smaller supremal values. The former case corresponds to **vague** or **underspecified concepts**, with positive and negative extensions that do not cover the domain (say for instance, music so confusing that it

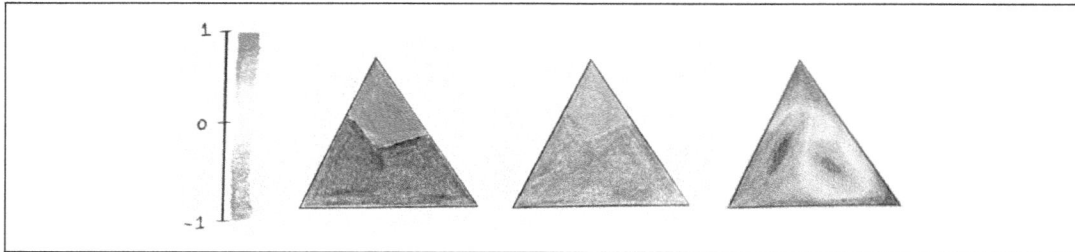

Figure 7: Various concepts over a simplex domain (say, that of Music, with vertices labelled {Metal,Jazz,Yodelling} according to taste.) Redder regions are positive extensions, and bluer regions are negative extensions. The leftmost concept is classical (and convex). The middle concept is uncertain: it never takes values $-1$ or $1$. The rightmost concept is vague: it has regions (yellow) in which it takes value 0.

cannot be classified into any genre known by the listener), and the latter case corresponds to **uncertain concepts**, which subsume the concepts that can be obtained by scaling classical concepts by an uncertainty parameter $0 < \gamma < 1$. Notably, these concepts arise from the scaling procedure when the resultant concept does not take on the full range of values $[-1, 1]$.

When $\epsilon \approx 1$, we force $\hat{f}$ to take value 0 almost everywhere in the domain, by a similar argument as in the case $\epsilon = 0$. These correspond to concepts with very sparse support in their domain, such as pointwise concepts. For example, a mechanism that detects the hex-colour #6600ff positively, while not responding when shown any other colour, would be an example of a such a pointwise concept functional with $\epsilon \approx 1$, that takes value 1 in a very small neighbourhood of #6600ff in the domain of colour, and 0 everywhere else.

Note that by construction, the normalised concepts $\hat{f}$ can never be such that $\|\hat{f}\| > 1$. Such bounding is only necessary if we are committed to the real interval $[-1, 1]$ representing fuzzy truth values, and may be dispensed with if we wish to treat our concepts as analogs of neural activation functions.

## 5 Modelling Negation and Similarity

In **Hilb**, we can already model concept-negation as the negative identity map, which maps $\hat{f} \mapsto -\hat{f}$, swapping the positive and negative extensions. This negation is involutive, but one can also have non-involutive negations, by for instance applying a scalar in $(0, 1)$ to $-\mathbb{I}_{L^2(\Delta)}$. The aforementioned cases subsume the expressive capacity of [14]. In our case, we may consider further generalised notions of negation via premultiplication by maps $\mathbb{N}$ which are "sufficiently close to $-\mathbb{I}_{L^2(\Delta)}$", where "sufficiently close" is expressed by a parameter $\delta \in [0,1]$:

$$\langle -\mathbb{I}_{L^2(\Delta)}, \mathbb{N} \rangle \leq \delta$$

Whenever $\Delta$ contains more than one point, $L^2(\Delta)$ is infinite-dimensional. Finite dimensional Hilbert spaces are isomorphic to their dual spaces, but this fails for infinite dimensional Hilbert spaces: the category **Hilb** of Hilbert spaces and bounded linear maps between them is not compact closed and is hence an unsuitable semantic category for Lambek pregroups, so we cannot place our $L^2(\Delta)$-spaces there.

We discuss several alternative methods of recovering the graphical tools of cups and caps in the conclusion. In this section, we demonstrate a particular method of relocation from **Hilb** to a variant of **LinRel** [21], the category of vector spaces and linear relations between them. In this structure, we win compatibility with categorical compositional semantics, and a parameterised similarity relation which subsumes identity, similarity, and negation.

## 5.1 The Category ConcFun

**Definition 5.1** (Linear Relations). *A linear relation between vector spaces $V$ and $W$ over the same field $F$ is a binary relation between their elements that:*

- Relates the zero vectors: $R(\mathbf{0}_V, \mathbf{0}_W)$.

- Is closed under vector addition: *for all $\mathbf{u}, \mathbf{v} \in V$ and all $\mathbf{w}, \mathbf{x} \in W$, if $R(\mathbf{u}, \mathbf{w})$ and $R(\mathbf{v}, \mathbf{x})$, then $R(\mathbf{u}+\mathbf{v}, \mathbf{w}+\mathbf{x})$.*

- Is closed under scalar multiplication: *for all $\mathbf{v} \in V$, all $\mathbf{w} \in W$, and all $\lambda \in F$, if $R(\mathbf{v}, \mathbf{w})$, then $R(\lambda \mathbf{v}, \lambda \mathbf{w})$.*

In short, Linear Relations between $V$ and $W$ relate linear subspaces of $V$ to linear subspaces of $W$. Observe that the definition of Linear Relations does not rely on the additive and multiplicative inverses present in the field $F$. We only wish to consider positive linear combinations of concepts in order to define a generalised negation later, so without compromising the spirit of Linear Relations, we can restrict the scalars to the addition-multiplication module over the positive reals.

Now we define the category of concept-functionals **ConcFun**, for which we will demonstrate a compact closed monoidal structure.

**Definition 5.2** (ConcFun). *The objects of **ConcFun** are the Hilbert Spaces with distinguished bases, where $\Delta$ are Borel unit-measure spaces, and the morphisms are linear relations between them over the $\mathbb{R}$-module $(\times, +, \mathbb{R}^{\geq 0})$.*

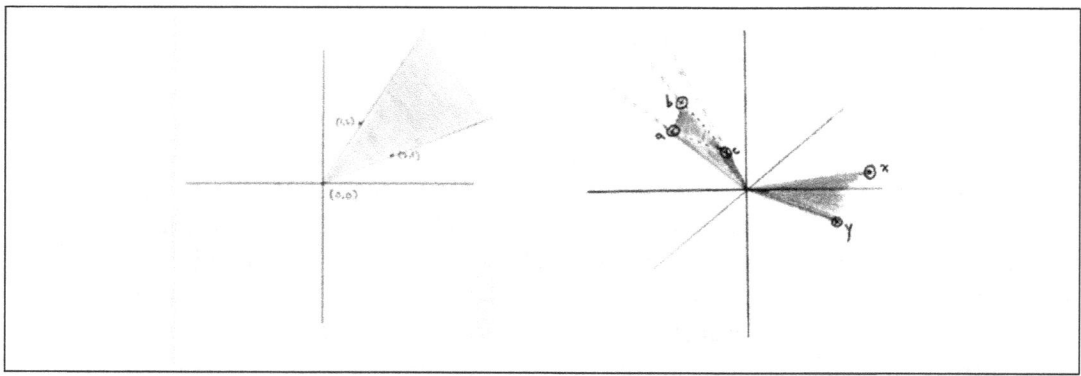

Figure 8: Some examples of how these modified linear relations behave. Left: The smallest linear relation $0 \to \mathbb{R}^2$ that contains the points $\{(1,2),(2,1)\}$. Right: A linear relation $\mathbb{R}^3 \to \mathbb{R}^3$ between three linearly independent points $\{a,b,c\}$ and $\{x,y\}$, which is the smallest that contains $\{(a,x),(b,x),(c,y)\}$; we show a cross section of the $\{a,b,c\}$-ray, and rays from the origin are related if they have the same colour. We might call these **Linear Raylations**.

**Proposition 5.1.** *ConcFun is a category.*

*Proof.* Every object of **ConcFun** is an object in **LinRel**, hence it will suffice to show that **ConcFun** is closed under relational composition. We must show relational closure under vector addition and scalar multiplication. Let $X \xrightarrow{R} Y \xrightarrow{S} Z$.

Vector Addition: Suppose $(S \circ R)(\mathbf{a}_1, \mathbf{c}_1)$ and $(S \circ R)(\mathbf{a}_2, \mathbf{c}_2)$. By relational composition, there exists $\mathbf{b}_1, \mathbf{b}_2$ such that $R(\mathbf{a}_1, \mathbf{b}_1), R(\mathbf{a}_2, \mathbf{b}_2), S(\mathbf{b}_1, \mathbf{c}_1), S(\mathbf{b}_2, \mathbf{c}_2)$.

By closure of $R$ and $S$ under vector addition, $R(\mathbf{a}_1 + \mathbf{a}_2, \mathbf{b}_1 + \mathbf{b}_2)$ and $S(\mathbf{b}_1 + \mathbf{b}_2, \mathbf{c}_1 + \mathbf{c}_2)$, and by relational composition, $(S \circ R)(\mathbf{a}_1 + \mathbf{a}_2, \mathbf{c}_1 + \mathbf{c}_2)$, as required.

Scalar Multiplication: $(S \circ R)(\mathbf{a}, \mathbf{c})$. By relational composition, there exists $\mathbf{b}$ such that $R(\mathbf{a}, \mathbf{b}), S(\mathbf{b}, \mathbf{c})$. By closure of $R$ and $S$ under scalar multiplication, for an arbitrary but fixed scalar $\gamma \in \mathbb{R}^{\geq 0}$, $R(\gamma\mathbf{a}, \gamma\mathbf{b})$ and $S(\gamma\mathbf{b}, \gamma\mathbf{c})$, so by relational composition, $(S \circ R)(\gamma\mathbf{a}, \gamma\mathbf{c})$, as required. □

In fact, every morphism in **LinRel** between objects of **ConcFun** is present in **ConcFun**: for an arbitrary morphism $X \xrightarrow{R} Y$ in **LinRel** between $X, Y$ objects of **ConcFun**, if $R(a,b)$, by closure under scalar multiplication in **LinRel**, $R(-a,-b)$, and there is a relation $X \xrightarrow{R'} Y$ in **ConcFun** that contains $(a,b)$ and $(-a,-b)$.

**ConcFun** is symmetric strict monoidal, being essentially the same as that of **LinRel**.

**Proposition 5.2.** *ConcFun is symmetric strict compact closed monoidal.*

*Proof.* Explicitly, the tensor product is the direct sum $\oplus$ of vector spaces (which we will write as column vectors), and the twist $\theta_{V,W}$ is the relation $\{(\begin{pmatrix}\mathbf{v}\\\mathbf{w}\end{pmatrix},\begin{pmatrix}\mathbf{w}\\\mathbf{v}\end{pmatrix}):\mathbf{v}\in V,\mathbf{w}\in W\}$. We identify singleton spaces in **ConcFun** that only contain the empty function $\emptyset \mapsto \mathbb{R}$ with the one-point zero subspace $\{\mathbf{0}\}$, which is the monoidal unit in **LinRel**. This unit $I$ is unique up to isomorphism, as there is a unique linear relation between the zero-subspaces of any two vector spaces. Hence **ConcFun** inherits the symmetric strict monoidal structure of **LinRel**.

It remains to demonstrate compact closure, which follows similarly to compact closure in **Rel**. For an arbitrary object $V$, we exhibit explicit self-dualities $\epsilon_V : V \oplus V \to I$, $\eta_V : I \to V \oplus V$ that satisfy yanking.

$$\epsilon_V := \{(\begin{pmatrix}\mathbf{v}\\\mathbf{v}\end{pmatrix}, \mathbf{0}) : \mathbf{v} \in V\} \qquad \eta_V := \{(\mathbf{0}, \begin{pmatrix}\mathbf{v}\\\mathbf{v}\end{pmatrix}) : \mathbf{v} \in V\}$$

From which it follows by relational composition that:

$$(\epsilon_V \oplus \mathrm{id}_V) \circ (\mathrm{id}_V \oplus \eta_V) = \mathrm{id}_V = (\mathrm{id}_V \oplus \epsilon_V) \circ (\eta_V \oplus \mathrm{id}_V)$$

□

The recovery of compact closure makes **ConcFun** a suitable semantics category for a Lambek Pregroup Grammar. Notably, all bounded linear maps in **Hilb** are subsumed by relations in **ConcFun** (as they are subsumed by linear relations in **LinRel**). In this new setting, we win a general, parameterised similarity relation which subsumes negation.

**Definition 5.3** ($\rho$-similarity in **ConcFun**). *For $\rho \in [-1, 1]$, and $V$ in **ConcFun**, define $\mathrm{Sim}_V^\rho : V \to V$ to be*

$$\mathrm{Sim}_V^\rho := \{(f, g) : g \in \mathbf{Ray}(\{h : \langle \hat{f}, h \rangle = \rho \text{ AND } h \text{ IS A CONCEPT}\})\}$$

*Where **Ray** is the linear span over the positive addition-multiplication $\mathbb{R}$-module ensuring that the relation is linear, and $f, g, h \in V$.*

Let us consider the action of $\mathrm{Sim}_V^\rho$ on the subspace of a single classical concept $f$, in the limit case when $\rho = 1$. The inner Ray is generated by the concepts $h$ such that $\langle f, g \rangle = 1$: *i.e.*, those classical concepts that agree with $f$ almost everywhere. So $\mathrm{Sim}_V^1$ behaves like $\mathrm{id}_V$. Symmetrically, $\mathrm{Sim}_V^{-1}$ behaves as the negation obtained from the negative-identity linear map (bottom left in Figure 9).

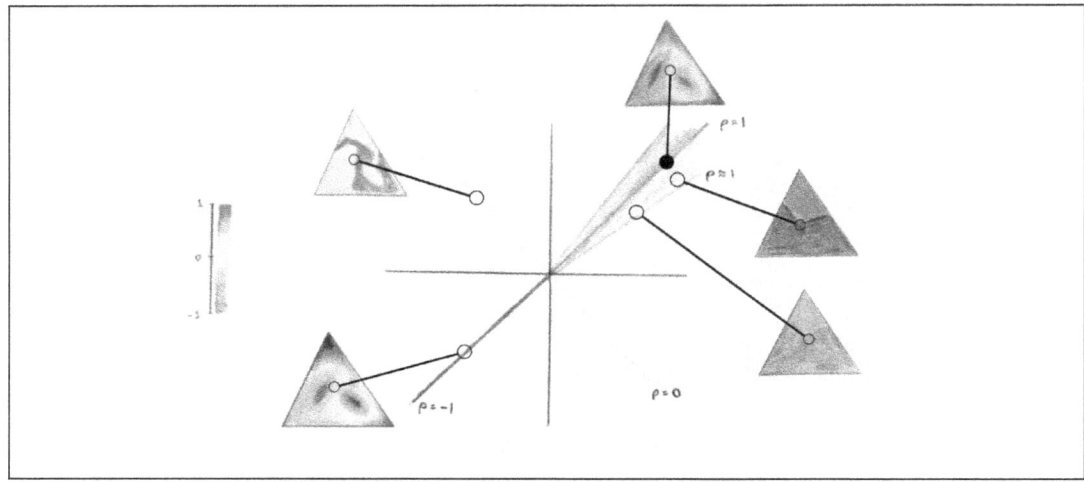

Figure 9: The action of $\text{Sim}^\rho_{L^2(\Delta)}$ on the subspace generated by a single concept (topmost, black circle), where $\Delta$ is the 2-simplex. Above is a visual representation of a 2D slice of the space $L^2(\Delta)$.

As $\rho$ decreases from 1, more concepts $h$ are introduced in the inner set, expanding the generated ray to include concepts of decreasing degrees of concordance with $f$ (right edge of Figure 9). Symmetrically, as $\rho$ increases from $-1$, the generated ray expands to include concepts of increasing degrees of concordance.

When $\rho = 0$, we relate $f$ to its orthogonal subspace. If $f$ has a penumbra, then notably, the image of $\text{Sim}^0_V$ under $f$ contains all those concepts that are defined in the penumbra, suggesting a form of concept-completion (upper left in Figure 9).

It is worth observing here a distinction in the literature on natural-language semantics of negation between "logical" and "pragmatic" [13] or "conversational" negation. We have been exclusively focused on the former, in which the negation of a concept is another algebraically-obtained concept. The latter form of pragmatic negation operates upon the universe of available predicates [15]: for instance, to infer that "The apple is maybe red" from "The apple is not green" requires a semantic-contextual restriction (driven by apple) to a space of compatible descriptors including the concepts red and green. While the similarity operator we have defined permits this form of alternative-capture, it only does so when all predicates involved share the same domain.

## 6 Treasure Hunt: spatial inference with negated and vague meanings

In this toy example, you are the captain of a crew of pirates seeking a buried treasure of precious gemstones, and you have found a treasure-map, narrowing the location of the treasure to a small region with two landmarks: a lake, and a palm tree. Tomorrow you must organise an expedition, but at the moment, you are drinking with an old man in a nearby village, hoping your charm can ply more information from him. Early in the evening, he tells you that "`The treasure is located west of the palm`". Much later, when he has finished all of your good rum, he tells you slurringly that "`The treasure is not located south of the lake`", before falling asleep.

We take the following grammatical types: $\{T$ (treasure), $S$ (search zone), $R$ (region), $L$ (landmark)$\}$. Our focus is the modelling of $R$ in **ConcFun** as $[-1, 1]$-valued functionals over the treasure-map viewed as a Borel unit-measure space; concepts in $R$ are regions on the map. $S$, the search zone, is similar to $R$, but may encode additional information relevant to your eventual decision, such as the topography of the region; the value of a functional on any point in $R$ corresponds to your prioritisation of searching that point. We can model $L$ as a regular vector space – such as the coordinates of the treasure-map – where the landmarks correspond to vectors in the space. It is unimportant how we model $T$. Let us focus on the first piece of information: "`The treasure is located west of the palm`". We assign words in our lexicon with types as follows: $\{$`the treasure`: $T$, `is`: $T^R S S^R T$, `located`: $T^R S R^L$, `west of`: $RL^L$, `the palm`: $L\}$. Figure 10 depicts the diagram of the sentence.

We model the determiner "`west of`" as a linear relation that relates landmarks to regions: concepts over a rectangular unit-measure space, depicted as the dotted rectangle surrounding the treasure map. The old man uses language as normal people do, so directional relations are not crisp concepts (see Figure above). We model "`located`" as concepts over the captain's internal representation of the region – depicted above as the treasure map.

Recall that $\text{Sim}^\rho$ behaves as the identity relation when $\rho = 1$, so in this case "`is`" is modelled as a copula, but we can model a spectra of such copula, where the $\rho$ parameter controls probity: for instance "`is almost certainly`" is obtained with large but less than unital $\rho$, while "`might be`" corresponds to smaller positive $\rho$, and "`is not`" corresponds to $\rho = -1$. We might take into account the drunkenness of the old man by setting the $\rho$ parameter to be negative but slightly greater than $-1$. So, we can visualise the second piece of information (collapsing the first few

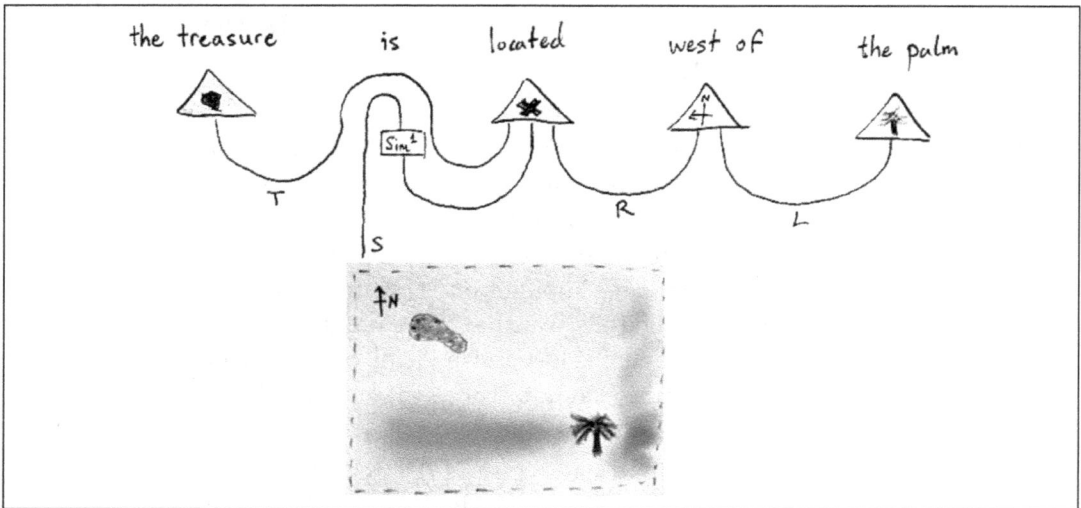

Figure 10: The composite state has potentially many functionals in its image. We provide a "typical representative" search zone concept among those functionals; redder regions on the treasure-map indicate areas of higher priority, and bluer regions indicate regions of lower priority.

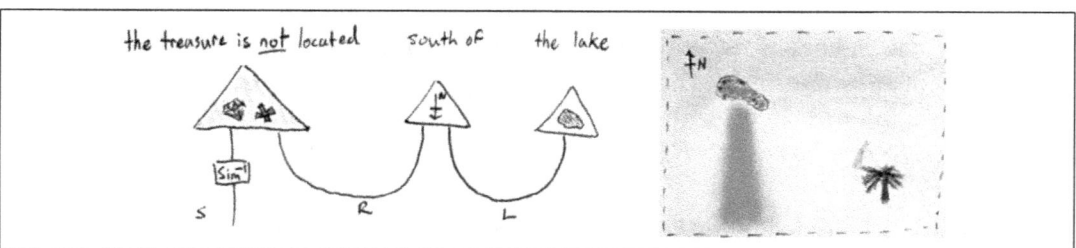

Figure 11: On the left, the diagram of the sentence. On the right, a typical representative concept from the composite state so obtained; note that this concept doesn't take on the value 1 anywhere, which may be the case, interpreting $-1 < \rho < 0$.

words graphically) as in (Figure 11.)

To put the information together, suppose that we type the word "and" as $S^L S S^R$, and model it as a relation from the tensor of two search zone concepts to a single one: a means to combine information. There is a great degree of modelling freedom here in the word "and", which we can exploit to model differing inferential strategies, as in Figure 13.

# CONCEPT FUNCTIONALS

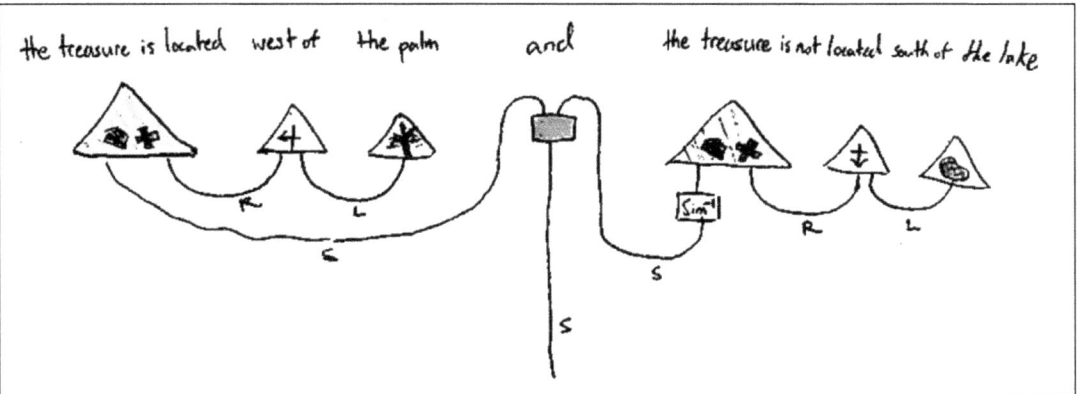

Figure 12: Treating the old man's information as a whole.

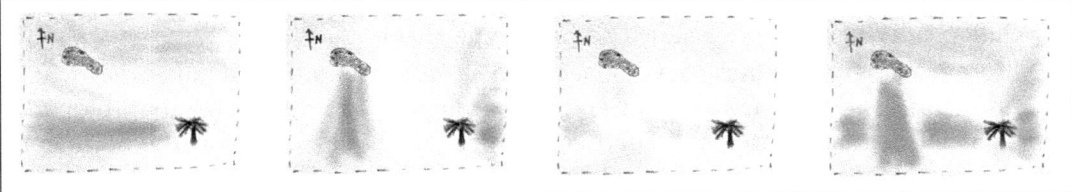

Figure 13: Typical concepts obtainable by taking the "and" of the concepts from Figures 10 and 11. The concept heatmaps are self-explanatory. From left to right, we depict typical concepts from the following "and" relations $S \otimes S \to S$ respectively: we give a generating set, from which one obtains the smallest linear raylation by closure under vector addition and scalar multiplication.

- $\{(\binom{f}{g}, [x \mapsto \max\{f(x), g(x)\}])\}$ ("greedy" pointwise maximisation)

- $\{(\binom{f}{g}, [x \mapsto \min\{f(x), g(x)\}])\}$ ("pessimistic" pointwise minimisation)

- $\{(\binom{f}{g}, \frac{f+g}{2})\}$ ("bayesian" convex combinations)

- $\{(\binom{f}{g}, [x \mapsto \underset{y \in \{f(x), g(x)\}}{\arg\max} \{\max\{y, -y\}\}])\}$ ("inferential" nonconvex combinations)

Your crew is looking quizzically over your shoulder at the strange diagrams you

741

have been drawing, and they have been for a while. You mutter "the concepts aren't crisp enough" (to general confusion), and you conclude that it is easier to just have the old man lead you to the treasure tomorrow himself, the least he could do for all of your good rum.

## 7 Conclusion

We have demonstrated how we may model and compute with a rich collection of concepts in **ConcFun**. Modelling concepts as functionals in this way provides room to express fuzzy, nonclassical, and probabilistic predicates, equipped with a parameterised notion of similarity that subsumes identity, similarity, and negation. There is room for future work to explore how concept functionals fit into the larger conceptual spaces framework, in particular bringing in quantitative representations for psychologically-based models, such as prototype theory.

Moreover, as we require our concepts to have measurable domains, we win a direct theoretical representation of spatial predicates, modularly compatable with differing inferential strategies. For future work, we hope to perform experiments upon linguistically disciplined models of machine spatial inference with this model. The linguistic discipline here is afforded by the categorical compositional approach, which we have shown here is compatible with our explicitly formulated, rather than distributionally obtained, semantics.

Regarding further practical application, allowing explicitly grounded semantics in conjunction with categorical grammar is potentially useful for the implementation of intelligent systems that must interface with reality beyond textual corpuses. The added value of perspicuously and explicitly defined rules of grammar lies in ruling out unwarranted behaviour in outcome-sensitive tasks. For instance, when an expensive robot with a camera is directed to navigate around "dangerous-looking things", for which it has a concept-recogniser. From work on distributional pragmatics of negation [15], it appears that the benefits of distributional and explicit semantics are complementary: distributionally obtained semantics encode relations between real-world concepts, while having explicit algorithms to recognise concepts "empirically" grounds the relata. For instance, we may use kernel density estimates obtained from instances of data as smooth concept functionals.

There are several ways to circumvent the obstacle of infinite-dimensional representations. Where the concepts in question may be approximated by models with finite parameters – as is the case for any particular trained Neural Net – we may treat each parameter as a basis vector, though notably at the price of linearity in the concepts expressed in the general case. Where the concept functionals themselves may be expressed by finite fourier approximants, linearity is conserved. Where the space of functionals carries differentiable structure, it becomes possible to recover stochastic gradient descent as a concrete method to learn concepts from data. There are, of course, alternative approaches. One is to use tools of [11] and [22] in which the compact closure requirement of categorical grammar is skipped entirely, permitting arbitrary semantic representations. Another is to use richer grammatical systems with structurally richer semantic categories, a more mature line of development for many extensions of the Lambek Calculus.

Theoretically, we have left much about **ConcFun** unexplored: to the best of our knowledge, the category of Linear "Raylations" is novel in this work, and merits further study as a middle case between the full unconstrained expressivity of the category of sets and relations **Rel**, and the relatively constrained linear structure of **Vect**, from which we may define notions of similarity from the inner product structure. Further, there is potential value in viewing negation as a byproduct of a parameterised similarity relation as we have defined, as over large enough or "universal" domains, we recover both the logical and pragmatic forms of negation, where the former asks for a sharper parameter of negation than the latter.

# References

[1] S. Abramsky and N. Tzevelekos. Introduction to categories and categorical logic. In B. Coecke, editor, *New Structures for Physics*, Lecture Notes in Physics, pages 3–94. Springer-Verlag, 2011.

[2] Y. Al-Mehairi, B. Coecke, and M. Lewis. Categorical compositional cognition. In *Quantum Interaction - 10th International Conference, QI 2016, San Francisco, CA, USA, July 20-22, 2016, Revised Selected Papers*, pages 122–134, 2016.

[3] J. Bolt, B. Coecke, F. Genovese, M. Lewis, D. Marsden, and R. Piedeleu. Interacting conceptual spaces. In *Semantic Spaces at the Intersection of NLP, Physics and Cognitive Science*, 2016.

[4] J. Bolt, B. Coecke, F. Genovese, M. Lewis, D. Marsden, and R. Piedeleu. Interacting conceptual spaces I. In M. Kaipainen, A. Hautamäki, P. Gärdenfors, and F. Zenker, editors, *Concepts and their Applications*, Synthese Library, Studies in Epistemology, Logic, Methodology, and Philosophy of Science. Springer, 2018. to appear.

[5] Marek Capinski and Peter E. Kopp. *Measure, Integral and Probability*. Springer, London ; New York, 2nd edition edition, May 2008.

[6] B. Coecke. The mathematics of text structure, 2019. arXiv:1904.03478.

[7] B. Coecke, F. Genovese, M. Lewis, and D. Marsden. Generalized relations in linguistics and cognition. In Juliette Kennedy and Ruy J. G. B. de Queiroz, editors, *Logic, Language, Information, and Computation - 24th International Workshop, WoLLIC 2017,*

London, UK, July 18-21, 2017, Proceedings, volume 10388 of *Lecture Notes in Computer Science*, pages 256–270. Springer, 2017.

[8] B. Coecke, E. Grefenstette, and M. Sadrzadeh. Lambek vs. Lambek: Functorial vector space semantics and string diagrams for Lambek calculus. *Annals of Pure and Applied Logic*, 164:1079–1100, 2013. arXiv:1302.0393.

[9] B. Coecke and A. Kissinger. *Picturing Quantum Processes. A First Course in Quantum Theory and Diagrammatic Reasoning*. Cambridge University Press, 2017.

[10] B. Coecke, M. Sadrzadeh, and S. Clark. Mathematical foundations for a compositional distributional model of meaning. In J. van Benthem, M. Moortgat, and W. Buszkowski, editors, *A Festschrift for Jim Lambek*, volume 36 of *Linguistic Analysis*, pages 345–384. 2010. arxiv:1003.4394.

[11] Antonin Delpeuch. Autonomization of Monoidal Categories. *arXiv:1411.3827 [cs, math]*, June 2019. arXiv: 1411.3827.

[12] P. Gärdenfors. *The Geometry of Meaning: Semantics Based on Conceptual Spaces*. MIT Press, 2014.

[13] P. Grice. *Studies in the Way of Words*. Harvard University Press, Cambridge, Mass., new ed edition edition, May 1991.

[14] Karl Moritz Hermann, Edward Grefenstette, and Phil Blunsom. âĂIJNot not badâĂİ is not âĂIJbadâĂİ: A distributional account of negation. page 9.

[15] Germán Kruszewski, Denis Paperno, Raffaella Bernardi, and Marco Baroni. There Is No Logical Negation Here, But There Are Alternatives: Modeling Conversational Negation with Distributional Semantics. *Computational Linguistics*, 42(4):637–660, December 2016.

[16] J. Lambek. The mathematics of sentence structure. *American Mathematics Monthly*, 65, 1958.

[17] J. Lambek. Type grammars as pregroups. *Grammars*, 4(1):21–39, 2001.

[18] G. Ludwig. *An Axiomatic Basis of Quantum Mechanics: Volume 1, Derivation of Hilbert Space*. Springer-Verlag, Berlin Heidelberg, 1985.

[19] P. Selinger. A survey of graphical languages for monoidal categories. In B. Coecke, editor, *New Structures for Physics*, Lecture Notes in Physics, pages 275–337. Springer-Verlag, 2011. arXiv:0908.3347.

[20] Lars Sivik and Charles Taft. Color naming: A mapping in the IMCS of common color terms. *Scandinavian Journal of Psychology*, 35(2):144–164, 1994.

[21] P. Sobocinski. Graphical linear algebra, 2015. http://graphicallinearalgebra.net.

[22] Vincent Wang. Graphical Grammar + Graphical Completion of Monoidal Categories. Master's thesis, University of Oxford.

[23] L. Zadeh. Fuzzy sets. *Information and Control*, 8(2205):339–353, 1965.

# Meaning Updating of Density Matrices

Bob Coecke
*Oxford University, Department of Computer Science*
*Cambridge Quantum Computing Ltd.*
`coecke@cs.ox.ac.uk`

Konstantinos Meichanetzidis
*Oxford University, Department of Computer Science*
*Cambridge Quantum Computing Ltd.*
`k.mei@cambridgequantum.com`

## Abstract

The DisCoCat model of natural language meaning assigns meaning to a sentence given: (i) the meanings of its words, and, (ii) its grammatical structure. The recently introduced DisCoCirc model extends this to text consisting of multiple sentences. While in DisCoCat all meanings are fixed, in DisCoCirc each sentence updates the meanings of words. In this paper we explore different update mechanisms for DisCoCirc, in the case where meaning is encoded in density matrices—which come with several advantages as compared to vectors.

Our starting point is two non-commutative update mechanisms, borrowing one from quantum foundations research [46, 47], and the other one that originally appeared in the area of our current interest, language meaning updating [15, 48]. Unfortunately, both of these lack key algebraic properties, nor are internal to the meaning category. Passing to double density matrices [3, 71] we do get an elegant internal diagrammatic update mechanism.

We also show that (commutative) spiders can be cast as an instance of the update mechanism of [46, 47]. This result is of interest to quantum foundations, as it bridges the work in Categorical Quantum Mechanics (CQM) with that on conditional quantum states. Our work also underpins the implementation of text-level Natural Language Processing (NLP) on quantum hardware, for which exponential space-gain and quadratic speed-up have previously been predicted.

We thank Jon Barrett, Stefano Gogioso and Rob Spekkens for inspiring discussions preceding the results in this paper, and Matt Pusey, Sina Salek and Sam Staton for comments on a previous version of the paper. KM was supported by the EPSRC National Hub in Networked Quantum Information Technologies.

# 1 Intro

Grammar is a mathematically well-studied structure [1, 44, 33, 45], and in Natural Language Processing (NLP) this mathematical structure is studied extensively in its own right, most notably for parsing. However, with respect to meanings of phrases and sentences it has not been given the respect it deserves, and some 10 years ago 'how to combine grammar and meaning' was still an open problem, and in particular, doing so in a conceptually well-founded manner.

The Categorical Distributional Semantics (DisCoCat) framework [28] was introduced in order to address this problem: it exploits grammatical structure in order to derive meanings of sentences from the meanings of its constituent words. For doing so we mostly relied on Lambek's pregroups [45], because of their simplicity, but any other mathematical model of grammar would work as well [20].

In NLP, meanings are established empirically (e.g. [35]), and this leads to a vector space representation. DisCoCat allows for meanings to be described in a variety of models, including the vector spaces widely used in NLP [28, 32, 41], but also relations as widely used in logic [28], density matrices [55, 4, 7], conceptual spaces [12], as well as many other more exotic models [52, 19].

Density matrices, which will be of interest to us in this paper, are to be conceived as an extension of the vector space model. Firstly, vector spaces do not allow for encoding lexical entailment structure such as in:

$$\texttt{tiger} \leq \texttt{big cat} \leq \texttt{mammal} \leq \texttt{vertebrate} \leq \texttt{animal}$$

while density matrices [66] do allow for this [4, 7]. Density matrices have also been used in DisCoCat to encode ambiguity (a.k.a. 'lack of information') [54, 40, 55]. Here the use of density matrices perfectly matches von Neumann's motivation to introduce them for quantum theory in the first place, and why they currently also underpin quantum information theory [9]. Density matrices also inherit the empirical benefits of vectors for NLP purposes. Other earlier uses of density matrices in NLP exploit the extended parameter space [11], which is a benefit we can also exploit.

DisCoCat does have some restrictions, however. It does not provide an obvious or unique mechanism for compositing sentences. Meanings in DisCoCat are also static, while on the other hand, in running text, meanings of words are subject to *update*, namely, the knowledge-update processes that the reader undergoes as they acquire more knowledge upon reading a text:

```
Once there was Bob.
Bob was a dog.
He was a bad dog that bites.
```

or, when properties of actors change as a story unfolds:

```
Alice and Bob were born.
    They got married.
   Then they broke up.
```

These restrictions of DisCoCat were addressed in the recently introduced DisCoCirc framework [15], in which sentences within larger text can be composed, and meanings are updated as text progresses. This raises the new question on what these update mechanisms are for specific models. Due to the above-stated motivations, we focus on meaning embeddings in density matrices.

There has been some use of meaning updating within DisCoCat, most notably, for encoding intersective adjectives [12] and relative pronouns [57, 58]. Here, a property is attributed to some noun by means of a suitable connective. Thus far, DisCoCat relied on the commutative special Frobenius algebras of CQM [27, 26], a.k.a. *spiders* [22, 23]. However, for the purpose of general meaning updating spiders are far too restrictive, for example, they force updating to be commutative. For this reason in this paper, we study several other update mechanisms and provide a unified picture of these, which also encompasses spiders.

In Section 3, we place meaning updating at the very centre of DisCoCirc: we show that DisCoCirc can be conceived as a theory about meaning updating only. This will in particular involve a representation of transitive verbs that emphasises how a verb creates a bond between the subject and the object. Such a representation has previously been used in [32, 42], where also experimental support was provided.

In Section 4 we identify two existing non-commutative update mechanisms for density matrices. The first one was introduced in [15, 48], which we will refer to as *fuzz*, and has a very clear conceptual grounding. The other one was introduced within the context of a quantum theory of Bayesian inference [46, 47], which we will refer to as *phaser*. While this update mechanism has been used in quantum foundations and has been proposed as a connective within DisCoCat [54], its conceptual status is much less clear, not in the least since it involves the somewhat ad hoc looking expression $\sqrt{\sigma} \rho \sqrt{\sigma}$ involving density matrices $\rho$ and $\sigma$. In Section 4.2 we show that, in fact, the phaser can be traced back to spiders, but in a manner that makes this update mechanism non-commutative. In Section 4.4 we point at already existing experimental evidence in favour of our update mechanisms.

In Section 5 we list a number of shortcomings of fuzz and phaser. Firstly, as we have two very distinct mechanisms, the manner in which meanings get updated is not unique. Both are moreover algebraically poor (e.g. they are non-associative). Finally, neither is internal to the meaning category of density matrices and CP-maps.

As both update mechanisms do have a natural place within a theory of meaning updating, in Section 6 we propose a mechanism that has fuzz and phaser as special cases. We achieve this by meanings and verbs as double density matrices [3, 71], which have a richer structure than density matrices. Doing so we still remain internal to the meaning category of density matrices and CP-maps, as we demonstrate in Section 7, where we also discuss implementation on quantum hardware.

In Section 8 we provide some very simple illustrative examples.

**Work related to DisCoCat and DisCoCirc.** Since DisCoCat came about around 2008 [13], there has been some related work, for example [8] soon after. Within the context of recurrent neural networks (RNNs) there also is some work now taking aspects of grammar into account, for example in [37, 62]. The paper [49] directly combines ideas from RNNs with DisCoCat, while aiming to maintain the richness and flexibility of the latter.

The sentence type used in this paper was also used in the recent DisCoCat paper that introduces Cartesian verbs [24], and as discussed in [24] Sec. 2.3, precursors of this idea are in [32, 43, 42]. Also within the context of DisCoCat, the work by Toumi et al. [16, 63] involves multi-sentence interaction by relying on *discourse representation structures* [39], which comes at the cost of reducing meaning to a number. Textual context is also present in the DisCoCat-related papers [56, 69], although no sentence composition mechanism is proposed.

Within more traditional natural language semantics research, *dynamic semantics* [34, 64] models sentence meanings as I/O-transitions and text as compositions thereof. However, the approach is still rooted in predicate logic, just as Montague semantics is, hence not accounting for more general meaning spaces, and also doesn't admit the explicit type structure of diagrams/monoidal categories. Dynamic semantics is a precursor of *dynamic epistemic logic (DEL)* [6, 5]; we expect that DEL, and generalisations thereof, may in fact emerge from our model of language meaning by considering an epistemic-oriented subset of meanings. In [59], static and dynamic vector meanings are explicitly distinguished, taking inspiration for the latter from dynamic semantics. There are many other logic-oriented approaches to text e.g. [2], of text organisation e.g. [51], and of the use of categorical structure.

# 2 Preliminaries

We expect the reader to have some familiarity with the DisCoCat framework [28, 57, 24], and with its diagrammatic formalism that we borrowed from Categorical Quantum Mechanics (CQM) [21, 22, 23], most notably caps/cups, spiders, and doubling.

We also expect the reader to be familiar with Dirac notation, projectors, density matrices, spectral decomposition and completely positive maps as used in quantum theory. We now set out the specific notational conventions that we will be following in this paper.

We read diagrams from top to bottom. Sometimes the boxes will represent linear maps, and sometimes they will represent completely positive maps. In order to distinguish these two representations we follow the conventions of [23, 22], which means that a vector and a density matrix will respectively be represented as:

$$\boxed{\text{vector}} \quad \longleftarrow \text{outputs} \longrightarrow \quad \boxed{\text{density}}$$

where wires represent systems. A privileged vector for two systems is the *cap*:

$$\frown \;:=\; \sum_i |ii\rangle$$

and its adjoint (a.k.a. 'bra') is the *cup*:

$$\smile \;:=\; \sum_i \langle ii|$$

Similarly to states, linear maps and CP maps are respectively depicted as:

$$\boxed{\text{linear}} \quad \longleftarrow \text{inputs} \longrightarrow \quad \boxed{\text{CP}}$$
$$\qquad\qquad \longleftarrow \text{outputs} \longrightarrow$$

We will reserve white dots to represent *spiders*:

$$\begin{array}{c}\cdots\\\circ\\\cdots\end{array} \;:=\; \sum_i |i\ldots i\rangle\langle i\ldots i| \qquad (1)$$

Crucially, spiders are clearly basis-dependent, and in fact, they represent orthonormal bases [27]. Note also that caps and cups are instances of spiders, and more generally, that spiders can be conceived as 'multi-wires' [23, 22]: the only thing that matters is what is connected to what by means of possibly multiple spiders, and not what the precise shape is of the spider-web that does so. This behaviour can be succinctly captured within the following *fusion* equation:

$$\text{(spider fusion diagram)} \;=\; \text{(single spider)} \qquad (2)$$

By *un-doubling* we refer to re-writing a CP-map as follows [60, 23]:

with the two boxes being $\sum_i f_i \otimes |i\rangle$ and $\sum_i |i\rangle \otimes \bar{f}_i$ for Kraus maps $f_i$, that is:

$$\boxed{\sum_i f_i \otimes |i\rangle} \quad \boxed{\sum_i |i\rangle \otimes \bar{f}_i} \tag{3}$$

In the specific case of density matrices this becomes:

$$\boxed{\text{dens}} \rightsquigarrow \boxed{\omega}\ \boxed{\omega} \tag{4}$$

Concretely, for a density matrix $\sum_i p_i |i\rangle\langle i|$ we have $\omega = \sum_i \sqrt{p_i}|ii\rangle$, i.e.:

$$\boxed{\sum_i \sqrt{p_i}|ii\rangle} \quad \boxed{\sum_i \sqrt{p_i}|ii\rangle}$$

## 3   Text meaning in DisCoCirc as updating

The starting point of both DisCoCat and DisCoCirc is the fact that pregroup analysis [45] of the grammatical structure associates to each sentences a diagram. In the case of a sentence (of which the associated grammatical type is denoted $s$) consisting of a subject (with type $n$ for 'noun'), a transitive verb (with composite type $^{-1}n \cdot s \cdot n^{-1}$), and an object (also with type $n$) this diagram looks as follows:

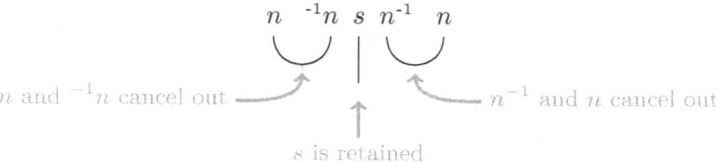

In DisCoCat [28] we then replace the types by the encoding of the word meanings,

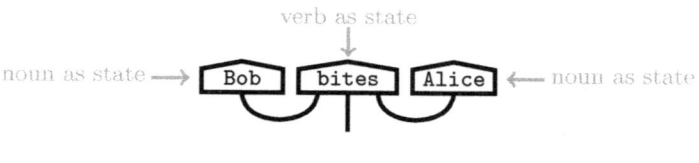

which in our case are represented by density matrices. The wires are interpreted as maps, for example, the cups will be the CP-maps associated to Bell-effects:

$$\sum_i \langle ii|$$

while the straight wire is an identity.

The above assumes that all words have fixed meanings given by those density matrices. However, as already explained in the introduction, meanings evolve in the course of developing text. Therefore, in DisCoCirc [15], prime actors like Bob and Alice in Bob bites Alice are not represented by a state, but instead by a wire carrying the evolving meaning. For this purpose we take the *s*-type to be the same as the type of that of the incoming actors [15]:

If we happen to have prior knowledge about that actor we can always plug a corresponding 'prior' state at the input of the wires:

which yields a DisCoCat-style picture. One can think of a sentence with an open input as a function $f$, while providing a prior state corresponds to $f$ being applied to a concrete input as in $f(x)$. The major advantage of not fixing states as a default is that this now allows us to compose sentences, for example:

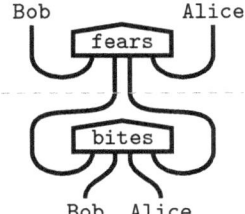

The above in particular means that the *s*-type will depend on the sentence. For example, in Bob is (a) dog, the noun dog can be taken to be fixed, so that the *s*-type becomes the Bob-wire alone, and we also introduce a special notation for is:

 (5)

Yanking wires this becomes:

$$\text{Bob} \quad \boxed{\text{dog}} \tag{6}$$

Here we think of dog as an adjective, for Bob. This reflects what we mean by an *update mechanism* in this paper: we update an actor's meaning (here Bob) by imposing a feature (here Dog) by means of the grey dot connective, where by 'actor' we refer to varying nouns subject to update.

We can also put more general transitive verbs into an adjective-like shape like in (6) by using the verb-form introduced in [32, 42]:

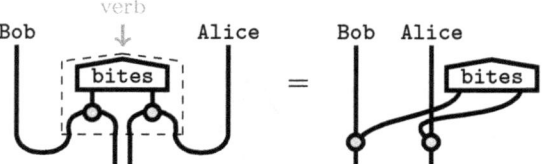

This representation of transitive verbs emphasises how a verb creates a bond between the subject and the object. From this point of view, an entire text can in principle be reduced to updates of this form, with the grey dot playing a central role.

The main remaining question now is:

*What is the grey dot?*

A first obvious candidate are the spiders (1), which have been previously employed in DisCoCat for intersective adjectives [12] and relative pronouns [57, 58]. However, while by spiders being multi-wires, by fusion (2) we have:

$$\curlyvee = \curlyvee$$

clearly we don't have:

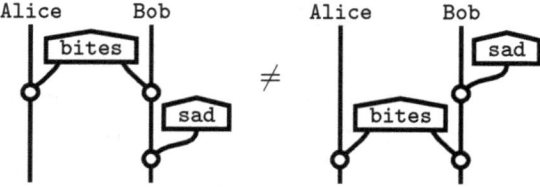

Therefore, in this case the grey dot needs to be something else. As spiders do make sense in certain cases, we desire something that has spiders as a special case, and as we shall see, this wish will be fulfilled.

## 4 Updating density matrices: fuzz vs. phaser

What is the most basic form of update for density matrices? Following Birkhoff-von Neumann quantum logic [10], any proposition about a physical system corresponds to a subspace $A$, or its corresponding projector $P_A$. Following [68, 67] it is also natural to think of propositions for natural language meaning like (being a) dog as such a projector. Imposing a proposition on a density matrix is then realised as follows:

$$P \circ - \circ P \qquad (7)$$

for example, $P_{\text{dog}} \circ \rho_{\text{Bob}} \circ P_{\text{dog}}$. Typically the resulting density matrix won't be normalised, so we will use the term density matrix also for sub-normalised and super-normalised positive matrices.

Now, by representing meanings by density matrices also dog itself would correspond to a density matrix in (6) for an appropriate choice of the grey dot. Fortunately, each projector $P$ is a (super-normalised) density matrix.

More generally, by means of weighted sums of projectors we obtain general density matrices in the form of their spectral decomposition:

$$\sum_i x_i P_i \qquad (8)$$

where one could imagine these sums to arise from the specific empirical procedure (e.g. [55, 50]) that is used to establish the meaning of dog. With dog itself a density matrix, we can now think of the grey dot in (6) as combining two density matrices:

$$\boxed{\text{Bob is (a) dog}} \ := \ \boxed{\text{Bob}} \ \boxed{\text{dog}} \qquad (9)$$

We now consider some candidates for such a grey dot. Firstly, let's eliminate two candidates. Composing two density matrices by matrix multiplication doesn't in general return a density matrix, nor does this have an operational interpretation. Alternatively, component-wise multiplication corresponds to fusion via spiders, which as discussed above is too specialised as it is commutative.

Two alternatives for these have already appeared in the literature:

$$\rho \,\fbox{}\, \sigma := \sum_i x_i \left( P_i \circ \rho \circ P_i \right) \qquad \text{with} \qquad \sigma := \sum_i x_i P_i \qquad (10)$$

$$\rho \,\fbox{}\, \sigma := \left( \sum_i x_i P_i \right) \circ \rho \circ \left( \sum_j x_j P_j \right) \qquad \text{with} \qquad \sigma := \sum_i x_i^2 P_i \qquad (11)$$

where for each we introduced a new dedicated 'guitar-pedal' notation. The first of these was proposed in [15, 48] specifically for NLP. The second one was proposed in the form $\sqrt{\sigma}\,\rho\,\sqrt{\sigma}$ within the context of a quantum theory of Bayesian inference [46, 47, 29]. Clearly, as update mechanisms, each of these can be seen as a quantitative generalisation of (7):

$$\sum_i x_i \left( P_i \circ - \circ P_i \right) \tag{12}$$

$$\left( \sum_i x_i P_i \right) \circ - \circ \left( \sum_j x_j P_j \right) \tag{13}$$

Diagrammatically, using un-doubling, we can represent the spectral decomposition of the density matrix $\sigma$ as follows:

and the fuzz and phaser seen as update mechanisms as in (6) then become:

where the state labeled $x$ is the vector $|x\rangle = (x_1 \ldots x_{n-1})^T$, the box labeled $P$ is the linear map $\sum_i P_i \otimes \langle i|$, and the white dot is a spider (1).

## 4.1 The fuzz

We call [symbol] the *fuzz*. The coefficients $x_i$ in the spectral decomposition of $\sigma$ are interpreted as representing the lack of knowledge about which proposition $P_i$ is imposed on $\rho$. In other words, the fuzz imposes a fuzzy proposition on the density matrix, and returns a density matrix comprising the mixture of having imposed different propositions each yielding a term $P_i \circ \rho \circ P_i$. This reflects the manner in which we would update a quantum state if there is uncertainty on which projector is imposed on a system undergoing measurement.

## 4.2 The phaser

We call ◉ the *phaser*. To understand the effect of the phaser, we write the 2nd argument $\sigma$ in (11) in terms of rank-1 projectors $P_i := |i\rangle\langle i|$ for an ONB $\{|i\rangle\}_i$, which can always be done by allowing some of the $x_i$'s to be the same. We have the following initial result relating the phaser to spiders:

**Lemma 4.1.** The phaser, when the 1st argument is pure, takes the form of a spider where the ONB in which the spider is expressed arises from diagonalisation of the 2nd argument. Setting $|x\rangle = (x_1 \ldots x_{n-1})^T$, we have:

$$\left(|\psi\rangle\langle\psi|\right) \circledcirc \left(\sum_i x_i^2 |i\rangle\langle i|\right) = |\phi\rangle\langle\phi|$$

where:

$$|\phi\rangle := \begin{array}{c}\boxed{\psi}\ \boxed{x}\\ \diagdown\diagup\\ |\end{array} \qquad (14)$$

So in particular, the resulting density matrix is also pure.

*Proof.* We have, using the fact the $x_i$'s are real:

$$\left(|\psi\rangle\langle\psi|\right) \circledcirc \left(\sum_i x_i^2 |i\rangle\langle i|\right) = \left(\sum_i x_i |i\rangle\langle i|\right) |\psi\rangle\langle\psi| \left(\sum_j x_j |j\rangle\langle j|\right)$$

$$= \left(\sum_i \langle i|\psi\rangle x_i |i\rangle\right) \left(\sum_j \langle\psi|j\rangle x_j \langle j|\right)$$

$$= \left(\sum_i \langle i|\psi\rangle x_i |i\rangle\right) \left(\sum_j \overline{\langle j|\psi\rangle} x_j \langle j|\right)$$

$$= \left(\sum_i \psi_i x_i |i\rangle\right) \left(\sum_j \bar{\psi}_i \bar{x}_j \langle j|\right)$$

with $\psi_i := \langle i|\psi\rangle$. As the explicit form of the spider is:

$$\begin{array}{c}\diagdown\diagup\\ |\end{array} = \sum_i |i\rangle\langle ii|$$

we indeed have:

$$\sum_i \psi_i x_i |i\rangle = |\phi\rangle = \begin{array}{c}\boxed{\psi}\ \boxed{x}\\ \diagdown\diagup\\ |\end{array}$$

what completes the proof. □

From this it now follows that the apparently obscure phaser, in particular due to the involvement of square root when presented as in [46, 47, 29], canonically generalises to the spiders previously used in DisCoCat:

**Theorem 4.2.** The action of the phaser on its first argument can be expressed in terms of spiders, explicitly, using the notations of (11), it takes the form:

$$-\; \bullet\; \sigma \;=\; \boxed{x} \qquad (15)$$

*Proof.* We have:

$$\boxed{x} \;=\; \left(\sum_j |j\rangle\langle jj|\right) \circ \left(1 \otimes \sum_i x_i |i\rangle\right) \;=\; \sum_i x_i |i\rangle\langle i|$$

which then yields the action of the phaser in the form (13). □

So in conclusion, the phaser boils down to the spiders that we are already familiarly with in DisCoCat, hence now solidly justifying its consideration by us in the first place. Moreover, there is one important qualification that will overcome our objection voiced above against using spiders given that they yield commutativity. Namely, these spiders may be expressed in different ONBs which they inherit from the 2nd argument $\sigma$, and if we update with nouns which diagonalise in different bases, then the corresponding spiders typically won't commute:

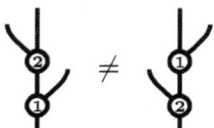

Hence, for the phaser, it is the properties with which the nouns are updated that control commutativity. The special case in which they commute is then the counterpart to the intersective adjectives [38] mentioned above.

Finally, we justify the term 'phaser'. Recalling that the key feature of the fuzz and the phaser is that they produce a density matrix, we see that we can let the $x_i$ in the phaser be complex:

$$\left(\sum_i x_i |i\rangle\langle i|\right) \circ - \circ \left(\sum_j \bar{x}_j |j\rangle\langle j|\right) \qquad (16)$$

In that case, of course, the density matrix $\sigma := \sum_i |x_i|^2 P_i$ does not fully specify (16), so rather than the density matrix, the data needed is the pair consisting of all $x_i$'s and $P_i$'s. Taking all $x_i$'s such that $|x_i| = 1$ then the operation:

takes the form of the original *phases* of ZX-calculus [17, 18, 23]. All spiders are equipped with phases, and more abstractly, they can be defined as certain Frobenius algebras [27]. In more recent versions of the ZX-calculus, more general phases are also allowed [53, 30], as these exist for equally general abstract reasons, and this then brings us to the general case of the phaser.

## 4.3 Normalisation for fuzz and phaser

We have the following no-go theorem for the fuzz:

**Proposition 4.3.** If the operation (12) sends normalised density matrices to normalised density matrices, then it must be equal to a (partial) decoherence operation:

$$\sum_i \Big( P_i \circ - \circ P_i \Big)$$

which retains all diagonal elements and sets off-diagonal ones to zero.

*Proof.* By trace preservation $\sum_i x_i (P_i \circ P_i) = \sum_i x_i P_i$ is the identity, so $x_i = 1$. □

For the phaser we have an even stronger result:

**Proposition 4.4.** If the operation (16) sends normalised density matrices to normalised density matrices, then for all $i$ we have $|x_i| = 1$. Taking the $x_i$'s to be positive reals, only the identity remains.

*Proof.* By trace preservation $(\sum_i x_i P_i) \circ (\sum_j \bar{x}_j P_j) = \sum_i x_i \bar{x}_i (P_i \circ P_i) = \sum_i |x_i|^2 P_i$ is the identity, so $|x_i| = 1$. □

It immediately follows from Propositions 4.3 and 4.4 that the operations (12) and (16) only preserve normalisation for a single trivial action both of ⧸ and ⊙. Of course, this was already the case for single projectors $P_A$, which will only preserve normalisation for fixed-points, so this result shouldn't come as a surprise. Hence, just like in quantum theory, one needs to re-normalise after each update if one insists on density matrices to be normalised.

## 4.4 Experimental evidence

Both the fuzz and the phaser have recently been numerically tested in their performance in modelling lexical entailment [48] (a.k.a. hyponymy). In [48] both fuzz and phaser are used to compose meanings of words in sentences, and it is explored how lexical entailment relationships propagate when doing so. The phaser performs particularly well, and seems to be very suitable when one considers more complex grammatical constructs. While these results were obtained within the context of DisCoCat, they also lift to the realm of DisCoCirc.

## 5 Non-uniqueness and non-internalness

The above poses a dilemma; there are two candidates for meaning update mechanisms in DisCoCirc. This seems to indicate that a DisCoCirc-formalism entirely based on updating, subject to that update process being unique, is not achievable.

Moreover, it is easy to check (and well-known for the phaser [36]) that both ![fuzz] and ![phaser] fail to have basic algebraic properties such as associativity, so treating them as algebraic connectives is not useful either. But that was never really our intention anyway, given that the formal framework where meanings in DisCoCat and DisCoCirc live is the theory of monoidal categories [25, 61]. In these categories, we both have states and processes which transform these states. In the case that states are vectors these process typically are linear maps, and in the case that states are density matrices these processes typically are CP-maps. However, neither the fuzz nor the phaser is a CP-map on the input $\rho \otimes \sigma$, which can clearly be seen from the double occurrence of projectors in their outputs. In other words, these update mechanisms are not *internal* to the meaning category. This means that there is no clear 'mathematical arena' where they live.

We will now move to a richer meaning category where the fuzz and phaser will be unified in a single construction which will become internal to the meaning category, as well as having a diagrammatic representation.

## 6 Pedalboard with double mixing

In order to unify fuzz ![fuzz] and phaser ![phaser], and also to make them internal to the meaning category, we use the *double density matrices* (DDMs) of [3, 71]. This is a new mathematical entity initially introduced within the context of NLP for capturing both lexical entailment and ambiguity within one structure [3]. On the other hand, they are a natural progression from the density matrices introduced by von Neumann for formulating quantum theory [65]. The key feature of DDMs for us is that they have two distinct modes of mixedness, for which we have the following:

**Theorem 6.1.** DDMs enable one to unify fuzz and phaser in a combined update mechanism, where fuzz and phaser correspond to the two modes of mixedness of DDMs. In order to do so, meanings of propositions are generalised to being DDMs, and the update dot is then entirely made up of wires only:

We now define DDMs, and continue with the proof of the theorem. Firstly, it is shown in [71] that there are two natural classes of DDMs, namely those arising from double dilation, and those arising from double mixing, and here we need the latter. While mixing can be thought of as passing from vectors to weighted sums of doubled vectors as follows (using the un-doubling representation):

$$|\phi\rangle \quad \rightsquigarrow \quad \sum_i x_i |\phi_i\rangle |\bar{\phi}_i\rangle$$

*double mixing* means repeating that process once more [3]:

$$\sum_i x_i |\phi_i\rangle |\bar{\phi}_i\rangle \quad \rightsquigarrow \quad \sum_{ijk} y_k x_{ik} x_{jk} |\phi_{ik}\rangle |\bar{\phi}_{ik}\rangle |\phi_{jk}\rangle |\bar{\phi}_{jk}\rangle$$

Setting $|\omega_{ik}\rangle := y_k^{1/4} x_{ik}^{1/2} |\phi_{ik}\rangle$ this becomes:

$$\sum_{ijk} |\omega_{ik}\rangle |\bar{\omega}_{ik}\rangle |\omega_{jk}\rangle |\bar{\omega}_{jk}\rangle$$

and in diagrammatic notation akin to that of un-doubled density matrices, we obtain the following generic form for double density matrices [71]:

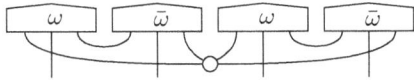

In order to relate DDMs to our discussion in Section 4, we turn them into CP-maps in the un-doubled from of (3), by bending up the inner wires:

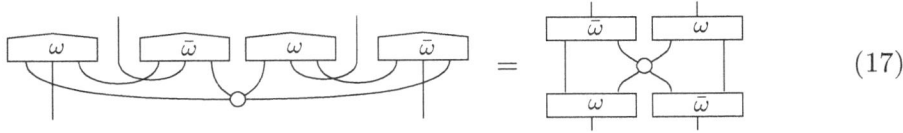 (17)

where:

as well as the horizontal reflections of these. In order to see that this is indeed an instance of (3), using fusion (2) we rewrite the spider in the RHS of (17) as follows:

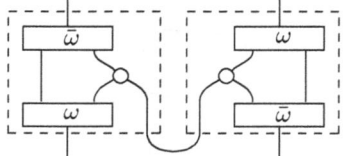

These CP-maps take the concrete form:

$$\sum_k \left( \sum_i |\omega_{ik}\rangle\langle\omega_{ik}| \right) \circ - \circ \left( \sum_j |\omega_{jk}\rangle\langle\omega_{jk}| \right) \qquad (18)$$

where the $k$-summation sums corresponds to the spider and the other summations to the two connecting wires. Using the spectral decomposition of the density matrix $\sum_i |\omega_{ik}\rangle\langle\omega_{ik}|$ this can be rewritten as follows, where the $y_k$'s are arbitrary:

$$\begin{aligned}
(18) &= \sum_k \left( \sum_i x'_{ik} P_{ik} \right) \circ - \circ \left( \sum_j x'_{jk} P_{jk} \right) \\
&= \sum_k y_k \left( \sum_i \frac{x'_{ik}}{y_k} P_{ik} \right) \circ - \circ \left( \sum_j \frac{x'_{jk}}{y_k} P_{jk} \right) \\
&= \sum_k y_k \left( \sum_i x_{ik} P_{ik} \right) \circ - \circ \left( \sum_j x_{jk} P_{jk} \right)
\end{aligned}$$

Now, this expression accommodates both the update mechanisms (12) and (13) as special cases, which are obtained by having either in the outer- or in the two inner-summations only a single index, and setting the corresponding scalar to 1. Consequently, in this form, we can think of the doubled density matrices as a canonical generalisation of propositions that unifies fuzz and phaser.

By relying on idempotence of the projectors we obtain:

$$(18) = \sum_k y_k \left( \sum_i x_{ik} P_{ik} \circ P_{ik} \right) \circ - \circ \left( \sum_j x_{jk} P_{jk} \circ P_{jk} \right)$$

and we can now indicate the roles of fuzz and phaser diagrammatically, as follows:

 (19)

where the summations and corresponding scalars are represented by their respective pedals. Of course, this notation is somewhat abusive, as also the projectors are part of the fuzz and phaser. Also, while the phaser appears twice in this picture, there is only one, just like for a density matrix $|\psi\rangle\langle\psi|$ there are two occurrences of $|\psi\rangle$.

We now put (19) in a form that exposes what the dot as in (6) is when taking meanings to be DDMs. Recalling the wire-bending we did in (17) we have:

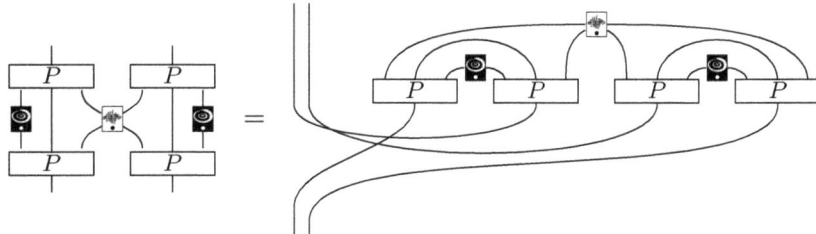

so it indeed follows that the dot only contains plain wires, which completes the proof.

**Remark 6.2.** One question that may arise concerns the relationship of the decomposition of CP-maps (19) and the Krauss decomposition of CP maps. A key difference is that (19) is a decomposition in terms of projections, involving two levels of sums, constituting it more refined or constrained than the more generic Krauss decomposition. This then leads to interesting questions, for example, regarding uniqueness of the projectors and coefficients arising from the spectral decompositions in our pedal-board construction. Furthermore, these coefficients may be used in a quantitative manner, e.g. for extracting entropies as in [55].

# 7 Meaning category and physical realisation

We can still take as meaning category density matrices with CP-maps as processes. We can indeed think of a DDM as a density matrix of the form (4):

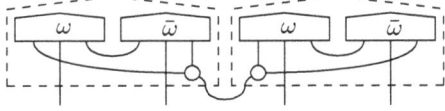

The update dot is a CP-map of the form (3):

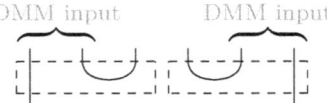

In this way we obtain a meaning category for which updating is entirely internal.

In previous work, we already indicated the potential benefits of the implementation of standard natural language processing tasks modelled in DisCoCat on a quantum computer [70]. One major upshot is the exponential space gain one obtains by encoding large vectors on many-qubit states. Another is the availability of quantum algorithms that yield quadratic speedup for tasks such as classification and question-answering. Moreover, the passage from vectors to density matrices is natural for a quantum computer, as any qubit can also be in a mixed state. Double density matrices are then implemented as mixed entangled states.

## 8 Some examples

**Example 1: Paint it black.** Following Theorem 4.2, the phaser ⊚ can be expressed as a spider-action (15). The aim of this example is to illustrate how in this form non-commutative updating arises. For the sake of clarity we will only consider rank-1 projectors rather than proper phasers, but this suffices for indicating how non-commutativity of update arises. The toy text for this example is:

```
Door turns red.
Door turns black.
```

Diagrammatically we have:

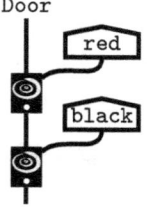

One can think of **turns** as an incarnation of **is** with non-commutative capability. Therefore it is of the form (5), and reduces to (6), where the grey dots are spiders. We take **red** and **black** to correspond to vectors $|r\rangle$ and $|b\rangle$, with induced density matrices $|r\rangle\langle r|$ and $|b\rangle\langle b|$. As they share the feature of both being colours, they are related (e.g. according to a factual corpus) and won't be orthogonal:

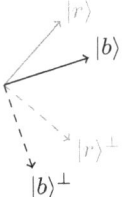

We can now build a ONB associated to `red`, with $|r\rangle$ one of the basis vectors, and the other ones taken from $|r\rangle^\perp$, which results in red spiders representing `red`, and then, taking $|x\rangle$ only to be non-zero for basis vector $|r\rangle$, `turns red` becomes:

Similarly, `turns black` becomes:

Crucially, the red and black spiders won't commute since they are defined on different ONBs taken from $\{|r\rangle, |r\rangle^\perp\}$ and $\{|b\rangle, |b\rangle^\perp\}$ respectively.

Concretely, we obtain two rank-1 projectors, $|r\rangle\langle r|$ and $|b\rangle\langle b|$ respectively, as desired. Taking an initial state for the door only considering the door's colour:

$$:= |d\rangle\langle d|$$

we obtain:

$$= \Big(|r\rangle\langle r|\Big) \circ \Big(|d\rangle\langle d|\Big) \circ \Big(|r\rangle\langle r|\Big) = z\,|r\rangle\langle r|$$

for some non-zero $z \in \mathbb{R}^+$, so now the door is `red`, and also:

$$= \Big(|b\rangle\langle b|\Big) \circ \Big(z\,|r\rangle\langle r|\Big) \circ \Big(|b\rangle\langle b|\Big) = z'\,|b\rangle\langle b|$$

so now the door is `black`, just like Mick Jagger wanted it to be.

**Example 2: Black fuzztones.** The aim of this example is to demonstrate the operation of the fuzz in modelling ambiguous adjectives. Above we treated `black` as pure, but in fact, it is ambiguous in that, for example, it may refer to a colour as well as to an art-gerne. This kind of ambiguity is accounted for by the fuzz. Disambiguation may take place when applying the ambiguous adjective to an appropriate noun, for example:

> `black poem`
> `black door`

or not, when the noun is lexically ambiguous as well, for example:

> `black metal`

which may be an art-genre, namely the music-genre, or the material:

This same ambiguity can propagate even further e.g.:

> `black metal fan`

which clearly demonstrates the importance of the fuzz, as ambiguity is ubiquitous in natural language. The latter example in fact also involves grammatical and syntactical ambiguity. This level of ambiguity is beyond the scope of this paper and will be studied elsewhere.

So now, besides $\texttt{black}_{col}$ as defined above for colour, there is another use of it, namely $\texttt{black}_{gen}$ for genre:

and we can represent the overall meaning as the following fuzz:

that is, concretely:

$$- \boxed{\text{\tiny{\textbf{\char"007E}}}}\; \sigma_{\texttt{black}} = y\left(|b_{col}\rangle\langle b_{col}| \circ - \circ |b_{col}\rangle\langle b_{col}|\right) + y'\left(|b_{gen}\rangle\langle b_{gen}| \circ - \circ |b_{gen}\rangle\langle b_{gen}|\right)$$

where the ambiguity is induced by the adjective:

$$\sigma_{\texttt{black}} := |b_{col}\rangle\langle b_{col}| + |b_{gen}\rangle\langle b_{gen}|$$

Inputting $\rho_{\text{poem}}$, $\rho_{\text{door}}$, and $\rho_{\text{metal}}$, we expect empirically (from a factual corpus) that the following terms will be very small:

$$\left(|b_{col}\rangle\langle b_{col}| \circ \rho_{\text{poem}} \circ |b_{col}\rangle\langle b_{col}|\right) \approx 0 \approx \left(|b_{gen}\rangle\langle b_{gen}| \circ \rho_{\text{door}} \circ |b_{gen}\rangle\langle b_{gen}|\right)$$

while these will all be significant:

$$\left(|b_{col}\rangle\langle b_{col}| \circ \rho_{\text{door}} \circ |b_{col}\rangle\langle b_{col}|\right) \gg 0 \qquad \left(|b_{gen}\rangle\langle b_{gen}| \circ \rho_{\text{poem}} \circ |b_{gen}\rangle\langle b_{gen}|\right) \gg 0$$

$$\left(|b_{col}\rangle\langle b_{col}| \circ \rho_{\text{metal}} \circ |b_{col}\rangle\langle b_{col}|\right) \gg 0 \qquad \left(|b_{gen}\rangle\langle b_{gen}| \circ \rho_{\text{metal}} \circ |b_{gen}\rangle\langle b_{gen}|\right) \gg 0$$

That is, `poem` and `door` are unambiguous nouns that disambiguate the ambiguous adjective `black`, while `metal` is ambiguous before and remains ambiguous after the application of the adjective `black` on it.

## 9 Outro

In this paper we proposed update mechanisms for DisCoCirc, in terms of fuzz and phaser, which in the real world look something like this:

and you can hear their sound in the QPL talk on the present paper [14]. Our pedal-notation has meanwhile also been adopted in [31], introducing the *compressor*.

We unified them within a single diagrammatically elegant update mechanism by employing double density matrices. In this way we upgraded the commutative spiders used in DisCoCat to non-commutative ones that respect the temporal order of the sentences within a text. The commutative spiders consist a special case of the more general phaser. At the same time, the fuzz models lexical ambiguity.

One might consider employing the double density matrix formalism to contribute to a theory of quantum Bayesian inference. Vice versa, fully incorporating inference within the diagrammatic formalism of quantum theory would aid in successfully modeling tasks in natural language processing as well as cognition.

Furthermore, since double density matrices can be described by standard density matrices with post-selection, the update formalism we have defined here can in principle be implemented in quantum hardware. Therefore, our framework provides a source for small- and large-scale experiments in the novel field of quantum natural language processing.

Finally, one might wonder whether an 'anatomy of completely positive maps' can be performed by means of double density matrices, potentially providing a compact framework in which to study quantum channels.

# References

[1] K. Ajdukiewicz. Die syntaktische konnexität. *Studia Philosophica*, 1:1–27, 1937.

[2] N. Asher and A. Lascarides. *Logics of conversation*. Cambridge University Press, 2003.

[3] D. Ashoush and B. Coecke. Dual Density Operators and Natural Language Meaning. *Electronic Proceedings in Theoretical Computer Science*, 221:1–10, 2016. arXiv:1608.01401.

[4] E. Balkir, M. Sadrzadeh, and B. Coecke. *Distributional Sentence Entailment Using Density Matrices*, pages 1–22. Springer International Publishing, Cham, 2016.

[5] A. Baltag, B. Coecke, and M. Sadrzadeh. Algebra and sequent calculus for epistemic actions. *Electronic Notes in Theoretical Computer Science*, 126:27–52, 2005.

[6] A. Baltag, L.S. Moss, and S. Solecki. The logic of public announcements, common knowledge, and private suspicions. In *Proceedings of the 7th conference on Theoretical aspects of rationality and knowledge*, pages 43–56. Morgan Kaufmann Publishers Inc., 1998.

[7] Dea Bankova, Bob Coecke, Martha Lewis, and Dan Marsden. Graded hyponymy for compositional distributional semantics. *Journal of Language Modelling*, 6(2):225–260, 2019.

[8] M. Baroni and R. Zamparelli. Nouns are vectors, adjectives are matrices: Representing adjective-noun constructions in semantic space. In *Proceedings of the 2010 conference on empirical methods in natural language processing*, pages 1183–1193, 2010.

[9] C. H. Bennett and P. W. Shor. Quantum information theory. *IEEE Trans. Inf. Theor.*, 44(6):2724–2742, September 2006.

[10] G. Birkhoff and J. von Neumann. The logic of quantum mechanics. *Annals of Mathematics*, 37:823–843, 1936.

[11] W. Blacoe, E. Kashefi, and M. Lapata. A quantum-theoretic approach to distributional semantics. In *Procs. of the 2013 Conf. of the North American Chapter of the Association for Computational Linguistics: Human Language Technologies*, pages 847–857, 2013.

[12] J. Bolt, B. Coecke, F. Genovese, M. Lewis, D. Marsden, and R. Piedeleu. Interacting conceptual spaces I: Grammatical composition of concepts. In M. Kaipainen, A. Hautamäki, P. Gärdenfors, and F. Zenker, editors, *Concepts and their Applications*, Synthese Library, Studies in Epistemology, Logic, Methodology, and Philosophy of Science. Springer, 2018. to appear.

[13] S. Clark, B. Coecke, and M. Sadrzadeh. A compositional distributional model of meaning. In *Proceedings of the Second Quantum Interaction Symposium (QI-2008)*, pages 133–140, 2008.

[14] B. Coecke. QPL 2020 Talk. https://www.youtube.com/watch?v=wP_Jsxn7BA4&t.

[15] B. Coecke. The mathematics of text structure, 2019. arXiv:1904.03478.

[16] B. Coecke, G. De Felice, D. Marsden, and A. Toumi. Towards compositional distributional discourse analysis. In M. Lewis, B. Coecke, J. Hedges, D. Kartsaklis, and D. Marsden, editors, Procs. of the 2018 Workshop on *Compositional Approaches in Physics, NLP, and Social Sciences*, volume 283 of *Electronic Proceedings in Theoretical Computer Science*, pages 1–12, 2018.

[17] B. Coecke and R. Duncan. Interacting quantum observables. In *Proceedings of the 37th International Colloquium on Automata, Languages and Programming (ICALP)*, Lecture Notes in Computer Science, 2008.

[18] B. Coecke and R. Duncan. Interacting quantum observables: categorical algebra and diagrammatics. *New Journal of Physics*, 13:043016, 2011. arXiv:quant-ph/09064725.

[19] B. Coecke, F. Genovese, M. Lewis, and D. Marsden. Generalized relations in linguistics and cognition. In Juliette Kennedy and Ruy J. G. B. de Queiroz, editors, *Logic, Language, Information, and Computation - 24th International Workshop, WoLLIC 2017, London, UK, July 18-21, 2017, Proceedings*, volume 10388 of *Lecture Notes in Computer Science*, pages 256–270. Springer, 2017.

[20] B. Coecke, E. Grefenstette, and M. Sadrzadeh. Lambek vs. Lambek: Functorial vector space semantics and string diagrams for Lambek calculus. *Annals of Pure and Applied Logic*, 164:1079–1100, 2013. arXiv:1302.0393.

[21] B. Coecke and A. Kissinger. Categorical quantum mechanics I: causal quantum processes. In E. Landry, editor, *Categories for the Working Philosopher*. Oxford University Press, 2016. arXiv:1510.05468.

[22] B. Coecke and A. Kissinger. Categorical quantum mechanics II: Classical-quantum interaction. *International Journal of Quantum Information*, 14(04):1640020, 2016.

[23] B. Coecke and A. Kissinger. *Picturing Quantum Processes. A First Course in Quantum Theory and Diagrammatic Reasoning*. Cambridge University Press, 2017.

[24] B. Coecke, M. Lewis, and D. Marsden. Internal wiring of cartesian verbs and prepositions. In M. Lewis, B. Coecke, J. Hedges, D. Kartsaklis, and D. Marsden, editors, Procs. of the 2018 Workshop on *Compositional Approaches in Physics, NLP, and Social*

*Sciences*, volume 283 of *Electronic Proceedings in Theoretical Computer Science*, pages 75–88, 2018.

[25] B. Coecke and É. O. Paquette. Categories for the practicing physicist. In B. Coecke, editor, *New Structures for Physics*, Lecture Notes in Physics, pages 167–271. Springer, 2011. arXiv:0905.3010.

[26] B. Coecke, É. O. Paquette, and D. Pavlović. Classical and quantum structuralism. In S. Gay and I. Mackie, editors, *Semantic Techniques in Quantum Computation*, pages 29–69. Cambridge University Press, 2010. arXiv:0904.1997.

[27] B. Coecke, D. Pavlović, and J. Vicary. A new description of orthogonal bases. *Mathematical Structures in Computer Science, to appear*, 23:555–567, 2013. arXiv:quant-ph/0810.1037.

[28] B. Coecke, M. Sadrzadeh, and S. Clark. Mathematical foundations for a compositional distributional model of meaning. In J. van Benthem, M. Moortgat, and W. Buszkowski, editors, *A Festschrift for Jim Lambek*, volume 36 of *Linguistic Analysis*, pages 345–384. 2010. arxiv:1003.4394.

[29] B. Coecke and R. W. Spekkens. Picturing classical and quantum bayesian inference. *Synthese*, 186(3):651–696, 2012.

[30] B. Coecke and Q. Wang. ZX-rules for 2-qubit clifford+T quantum circuits. In Jarkko Kari and Irek Ulidowski, editors, *Reversible Computation - 10th International Conference, RC 2018, Leicester, UK, September 12-14, 2018, Proceedings*, volume 11106 of *Lecture Notes in Computer Science*, pages 144–161. Springer, 2018.

[31] G. De las Cuevas, A. Klingler, M. Lewis, and T. Netzer. Cats climb entails mammals move: preserving hyponymy in compositional distributional semantics. *arXiv:2005.14134 [quant-ph]*, 2020.

[32] E. Grefenstette and M. Sadrzadeh. Experimental support for a categorical compositional distributional model of meaning. In *The 2014 Conference on Empirical Methods on Natural Language Processing.*, pages 1394–1404, 2011. arXiv:1106.4058.

[33] V.N. Grishin. On a generalization of the Ajdukiewicz-Lambek system. In *Studies in nonclassical logics and formal systems*, pages 315–334. Nauka, Moscow, 1983.

[34] J. Groenendijk and M. Stokhof. Dynamic predicate logic. *Linguistics and philosophy*, 14(1):39–100, 1991.

[35] Z. S. Harris. Distributional structure. *Word*, 10(2-3):146–162, 1954.

[36] D. Horsman, C. Heunen, M. F. Pusey, J. Barrett, and R. W. Spekkens. Can a quantum state over time resemble a quantum state at a single time? *Proceedings of the Royal Society A: Mathematical, Physical and Engineering Sciences*, 473(2205):20170395, 2017.

[37] O. Irsoy and C. Cardie. Deep recursive neural networks for compositionality in language. In *Advances in neural information processing systems*, pages 2096–2104, 2014.

[38] H. Kamp and B. Partee. Prototype theory and compositionality. *Cognition*, 57:129–191, 1995.

[39] H. Kamp and U. Reyle. *From discourse to logic: Introduction to modeltheoretic semantics of natural language, formal logic and discourse representation theory*, volume 42.

Springer Science & Business Media, 2013.

[40] D. Kartsaklis. *Compositional Distributional Semantics with Compact Closed Categories and Frobenius Algebras*. PhD thesis, University of Oxford, 2014.

[41] D. Kartsaklis and M. Sadrzadeh. Prior disambiguation of word tensors for constructing sentence vectors. In *The 2013 Conference on Empirical Methods on Natural Language Processing.*, pages 1590–1601. ACL, 2013.

[42] D. Kartsaklis and M. Sadrzadeh. A study of entanglement in a categorical framework of natural language. In *Proceedings of the 11th Workshop on Quantum Physics and Logic (QPL)*. Kyoto âĂŽJapan, 2014.

[43] D. Kartsaklis, M. Sadrzadeh, S. Pulman, and B. Coecke. Reasoning about meaning in natural language with compact closed categories and Frobenius algebras. In *Logic and Algebraic Structures in Quantum Computing and Information*. Cambridge University Press, 2015. arXiv:1401.5980.

[44] J. Lambek. The mathematics of sentence structure. *American Mathematics Monthly*, 65, 1958.

[45] J. Lambek. From word to sentence. *Polimetrica, Milan*, 2008.

[46] M. S. Leifer and D. Poulin. Quantum graphical models and belief propagation. *Annals of Physics*, 323(8):1899–1946, 2008.

[47] M. S. Leifer and R. W. Spekkens. Towards a formulation of quantum theory as a causally neutral theory of bayesian inference. *Physical Review A*, 88(5):052130, 2013.

[48] M. Lewis. Compositional hyponymy with positive operators. In *Proceedings of Recent Advances in Natural Language Processing, Varna, Bulgaria*, pages 638–647, 2019.

[49] M. Lewis. Compositionality for recursive neural networks. *arXiv:1901.10723*, 2019.

[50] M. Lewis. Towards negation in DisCoCat, 2019. Proceedings of SemSpace 2019.

[51] W. C. Mann and S. A. Thompson. Rhetorical structure theory: Toward a functional theory of text organization. *Text-Interdisciplinary Journal for the Study of Discourse*, 8(3):243–281, 1988.

[52] D. Marsden and F. Genovese. Custom hypergraph categories via generalized relations. In *7th Conference on Algebra and Coalgebra in Computer Science (CALCO 2017)*. Schloss Dagstuhl-Leibniz-Zentrum fuer Informatik, 2017.

[53] K. F. Ng and Q. Wang. Completeness of the zx-calculus for pure qubit clifford+ t quantum mechanics. *arXiv preprint arXiv:1801.07993*, 2018.

[54] R. Piedeleu. Ambiguity in categorical models of meaning. Master's thesis, University of Oxford, 2014.

[55] R. Piedeleu, D. Kartsaklis, B. Coecke, and M. Sadrzadeh. Open system categorical quantum semantics in natural language processing. In *6th Conference on Algebra and Coalgebra in Computer Science (CALCO 2015)*. Schloss Dagstuhl-Leibniz-Zentrum fuer Informatik, 2015.

[56] T. Polajnar, L. Rimell, and S. Clark. An exploration of discourse-based sentence spaces for compositional distributional semantics. In *Proceedings of the First Workshop on Linking Computational Models of Lexical, Sentential and Discourse-level Semantics*,

pages 1–11, 2015.

[57] M. Sadrzadeh, S. Clark, and B. Coecke. The Frobenius anatomy of word meanings I: subject and object relative pronouns. *Journal of Logic and Computation*, 23:1293–1317, 2013. arXiv:1404.5278.

[58] M. Sadrzadeh, S. Clark, and B. Coecke. The Frobenius anatomy of word meanings II: possessive relative pronouns. *Journal of Logic and Computation*, 26:785–815, 2016. arXiv:1406.4690.

[59] M. Sadrzadeh and R. Muskens. Static and dynamic vector semantics for lambda calculus models of natural language. *arXiv:1810.11351*, 2018.

[60] P. Selinger. Dagger compact closed categories and completely positive maps. *Electronic Notes in Theoretical Computer Science*, 170:139–163, 2007.

[61] P. Selinger. A survey of graphical languages for monoidal categories. In B. Coecke, editor, *New Structures for Physics*, Lecture Notes in Physics, pages 275–337. Springer-Verlag, 2011. arXiv:0908.3347.

[62] K. S. Tai, R. Socher, and C. D. Manning. Improved semantic representations from tree-structured long short-term memory networks. *arXiv preprint arXiv:1503.00075*, 2015.

[63] A. Toumi. Categorical compositional distributional questions, answers & discourse analysis. Master's thesis, University of Oxford, 2018.

[64] A. Visser. Contexts in dynamic predicate logic. *Journal of Logic, Language and Information*, 7(1):21–52, 1998.

[65] J. von Neumann. Wahrscheinlichkeitstheoretischer aufbau der quantenmechanik. *Nachrichten von der Gesellschaft der Wissenschaften zu Göttingen, Mathematisch-Physikalische Klasse*, 1:245–272, 1927.

[66] J. von Neumann. *Mathematische grundlagen der quantenmechanik*. Springer-Verlag, 1932. Translation, *Mathematical foundations of quantum mechanics*, Princeton University Press, 1955.

[67] D. Widdows. Orthogonal negation in vector spaces for modelling word-meanings and document retrieval. In *41st Annual Meeting of the Association for Computational Linguistics*, Japan, 2003.

[68] D. Widdows and S. Peters. Word vectors and quantum logic: Experiments with negation and disjunction. *Mathematics of language*, 8(141-154), 2003.

[69] G. Wijnholds and M. Sadrzadeh. Classical copying versus quantum entanglement in natural language: The case of vp-ellipsis. *arXiv:1811.03276*, 2018.

[70] W. Zeng and B. Coecke. Quantum algorithms for compositional natural language processing. *Electronic Proceedings in Theoretical Computer Science*, 221, 2016. arXiv:1608.01406.

[71] M. Zwart and B. Coecke. Double Dilation $\neq$ Double Mixing. *Electronic Proceedings in Theoretical Computer Science*, 266:133–146, 2018. arXiv:1704.02309.

# Towards Logical Negation in Compositional Distributional Semantics

Martha Lewis*
*Dept. of Engineering Mathematics, University of Bristol*
*ILLC, University of Amsterdam*
martha.lewis@bristol.ac.uk

## Abstract

The categorical compositional distributional model of meaning gives the composition of words into phrases and sentences pride of place. However, it has so far lacked a model of logical negation. This paper gives some steps towards providing this operator, modelling it as a version of projection onto the subspace orthogonal to a word. We give a small demonstration of the operator's performance in a sentence entailment task.

## 1 Introduction

Compositional models of meaning aim to represent the meaning of phrases and sentences by combining representations of the words in the sentence according to some rule. Compositional distributional models, such as described in [3, 12, 30] combine the compositional approach with vector-based models of word meaning. In these models, nouns are represented as vectors, and function words, such as verbs and adjectives, are modelled as linear maps. In this paper, we use the *categorical compositional distributional (DisCoCat)* model introduced in [12]. This model formalises the compositional approach to language using category theory, setting up a functorial mapping between the grammar of the language on the one hand, and the structures used to represent lexical meaning on the other. In modelling the meaning of words and sentences, a distinction can be made between words with lexical content, and words that can arguably be modelled as an operation on the structure of the sentence. For example, in [36], relative pronouns are modelled as routing information around a sentence using the structure of a Frobenius algebra. In [19],

---

Many thanks to the reviewers for useful comments, guidance and recommendations
*This work is supported by NWO Veni project 'Metaphorical Meanings for Artificial Agents'.

conjunctions are modelled using Frobenius algebras. In [12], negation is modelled as a linear map on a two-dimensional sentence space which sends each basis vector to the subspace orthogonal to it. This idea of modelling negation as projection to the orthogonal subspace was used in [44], but at the vector level is somewhat unsatisfactory since a word and its negation are then of two different kinds. Furthermore within DisCoCat, words should be modelled as linear maps, which projection onto an orthogonal subspace doesn't satisfy.

There are various kinds of negation. Modelling negation as projection to an orthogonal subspace, as described above, has the result that the negated word or sentence is not similar to the word itself. However, it has also been argued that the negation of a word *should* be fairly similar to the original. In [21] the notion of *conversational negation* is introduced, in which the negation of a word should be viewed as introducing a range of alternatives to the negated concept. For example, *That's not a horse, it's a donkey* makes sense, whereas *That's not a horse, it's a score* does not. It is difficult to imagine a natural context where this sentence would be uttered. Amongst other findings, they show that pairs of items that are close together in a semantic space form good alternatives within this kind of sentence. For example, they have pairs such as *lizard/iguana* or *trumpet/saxophone*. Similarly, [17] argue that 'not red' is still a colour, and provide a model where the vector is divided into domains and only part of the vector is inverted. In [34] negation is viewed as antonymy, and they provide a model of negation in which an encoder is trained to produce the antonym of a given adjective.

In [44] the aim is not to model conversational negation, but to model negation within the context of information retrieval. We take a similar stance. Whilst distributional semantics is about modelling human use of language, as derived from text corpora, there is also a range of research into distributional representations for natural language inference, some incorporating logical semantics [4, 43, 8]. In this work, we aim to give an account of negation that can be used for natural language inference. In previous work [25, 24], we looked at hyponymy relations between individual words, and how that relation lifts to entailment between short sentences. In this work, we will introduce a way of modelling negation that interacts with our notions of hyponymy and composition in a way that allows us to model entailment between sentences. As such, the operator we present is more similar to a notion of negation then the pragmatic, conversational, notion described in [21].

Within the categorical compositional framework, we can be flexible about how to represent word meanings. In [12] the category **FVect** of vector spaces and linear maps was used, meaning that nouns and sentences are modelled as vectors, and functional words such as verbs and adjectives are modelled as linear maps. In [7] the category **ConvexRel** was used, enabling the representation of nouns and sentences as convex sets and function words as convex relations. In this paper we will use the category **CPM(FVect)** which models nouns and sentences as positive operators, and function words as completely positive maps. This approach to meaning was developed in [1, 2] and implemented in [24, 25]. We model negation as an operation related to projection onto the orthogonal subspace. In the current paper we will concentrate on negating nouns and verbs, however, the proposal we make can be applied to other forms such as adjectives, adverbs, or verb phrases. Algebraically, the operation we propose can also be applied to sentences. However, more work is needed to determine how this makes sense linguistically, since we do not currently have an account of quantification in this framework (but see [16] for an alternative account).

## 2 Background

### 2.1 Categorical compositional approaches to meaning

The categorical compositional model of meaning uses the framework of category theory to set up a mapping between the grammar of a language and the structures used to represent the manings of individual words. A formalization of grammar is chosen, and represented as a category, called the *grammar category*. A choice is made about the type of meaning representation, which again is formalized as a category, called the *meaning category*. The meaning category and the grammar category are chosen to have the same abstract structure. Type reductions in the grammar category are then functorially mapped to operations in the semantics category. In this paper, the grammar category and the meaning category are both *compact closed*. For details of what this means within the context of linguistics, see [12] or [33]. A gentle presentation is also given in [7].

**Pregroup grammar** In this paper, we will use pregroup grammar, although the formalism is flexible about what can be used, and other choices are given in, for example, [10, 26, 29]. A pregroup is a partially ordered monoid $(X, \cdot, 1, \leq)$ where each $x \in X$ has a left and a right adjoint $(-)^l, (-)^r$ such that:

$$\epsilon_x^r : x \cdot x^r \leq 1, \quad \epsilon_x^l : x^l \cdot x \leq 1 \quad \eta_x^r : 1 \leq x^r \cdot x, \quad \eta_x^l : 1 \leq x \cdot x^l \quad (1)$$

A pregroup grammar is the pregroup freely generated over a set of chosen types. We consider the set containing $n$ for noun and $s$ for sentence. Complex types are built up by concatenation of types, and we often leave out the dot so that $xy = x \cdot y$. If $x \leq y$ we say that $x$ *reduces to* $y$.

A string of types $t_1, ... t_n$ is *grammatical* if it reduces, via the morphisms above, to the sentence type $s$. For example, typing *clowns* as $n$, *tell* as $n^r s n^l$ and *the truth* as $n$, the sentence *Clowns tell the truth* has type $n(n^r s n^l)n$ and is shown to be grammatical as follows:

$$(\epsilon^r\ 1\ \epsilon^l)n(n^r s n^l)n \leq (\epsilon^r\ 1)(n\ n^r s\ 1) \tag{2}$$
$$\leq 1\ s\ 1 = s \tag{3}$$

The above reduction can be represented graphically as follows:

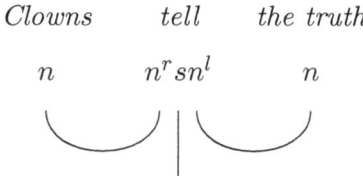

**Meaning categories** As a first example we describe how pregroup grammar is mapped to **FVect**, the category of vector spaces and linear transformations. The noun type $n$ is mapped to a vector space $N$ and the sentence type $s$ to $S$. The concatenation operation in the grammar is mapped to $\otimes$, i.e., the tensor product of vector spaces. Then the morphisms $\epsilon_x^r$ and $\epsilon_x^l$ map to tensor contraction, and $\eta_x^r$ and $\eta_x^l$ map to identity matrices.

Function words like verbs and adjectives are modelled as (multi)linear maps. Intransitive verbs are represented as maps from $N$ to $S$, or matrices in $N \otimes S$, and transitive verbs are represented as maps from two copies of $N$ to $S$, or tensors in $N \otimes S \otimes N$. So, in the example above, *Clowns* is mapped to a vector in $N$, as is *the truth*, and *tell* is mapped to a tensor in $N \otimes S \otimes N$. The vectors and tensors are concatenated using the tensor product, and tensor contraction is applied to map the sentence down into one sentence vector.

Compact closed categories have a nice diagrammatic calculus, described in [39], or for a linguistically couched explanation see [12]. In this calculus, the composition of the words *Clowns*, *tell*, and *the truth* into the sentence *Clowns tell the truth* is expressed as follows:

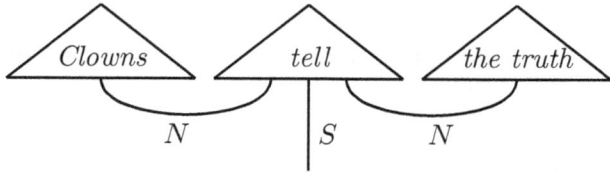

We will use this notation later to describe how to build particular representations of verbs and other function words.

In [12], the authors model negation as an operator on the sentence space. However, they only give a semantics for that operator in the case where the sentence space is two dimensional, and the dimensions correspond to truth and falsity - then, negation can be modelled as a swap operator, taking true to false and vice versa. In that model, the sentence *Clowns do not tell the truth* is given by the following diagram:

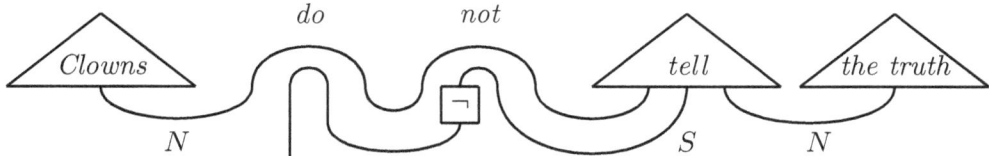

which can be simplified to:

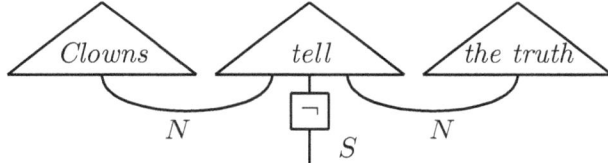

In this paper, we will take a different view, in which negation is applied at the word level, rather than the sentence level. This is because, as mentioned, we do not currently have a model of quantification, making negation at the sentence level harder to deal with.

## 2.2 Modelling words as positive operators

In [32], [2], and [1] the DisCoCat model is instantiated with the meaning category **CPM(FVect)**. This has the same objects as **FVect**, but the morphisms are now completely positive maps. The **CPM** construction is introduced in [38]. Words are now represented as positive operators rather than as vectors, and maps between them are completely positive maps. In the rest of the paper, we will use bra-ket notation which can be explained simply as follows. Suppose we are in a vector space $V = \mathbb{R}^n$, which is usually the case in distributional semantic applications. Then:

- A ket $|v\rangle \in V$ is a column vector.
- A bra $\langle v| \in V$ is a row vector.

- $\langle v | w \rangle$ is the inner product of $|v\rangle$ and $|w\rangle$
- $|v\rangle \langle w|$ is the outer product of $|v\rangle$ and $|w\rangle$

A positive operator is then defined as follows. For a unit vector $|v\rangle$, the projection operator $|v\rangle \langle v|$ onto the subspace spanned by $|v\rangle$ is called a *pure state*. A positive operator is given by sum of pure states. It is an operator $A$ such that:

1. $\forall v \in V. \langle v|A|v\rangle \geq 0$,
2. $A$ is self-adjoint

If, in addition, $A$ has trace 1, then $A$ encodes a probabilistic mixture of pure states, and is called a density matrix. Relaxing this condition gives us different choices for normalization.

We give an informal description of how pregroup grammar maps into the category **CPM(FVect)**. For more details see [32], [2], or [1]. Within **CPM(FVect)**, the objects are vector spaces, and morphisms are completely positive maps, i.e. linear maps that preserve positivity of operators and do so for any trivial extension. The underlying spaces in which we represent nouns, sentences, and other words in are now doubled up, meaning that a noun is a positive operator $N \to N$, or a positive semidefinite matrix in $N^* \otimes N$. Completely positive maps are defined in [38] as a morphism $\phi : A^* \otimes A \to B^* \otimes B$ such that there exists an object $C$ in the underlying category, in our case **FVect**, and a morphism $k : C \otimes A \to B$ such that:

$$\phi = (k_* \otimes k) \circ (1_{A^*} \otimes \eta_C \otimes 1_A)$$

Importantly, **CPM(FVect)** is also compact closed, so that the same sort of functorial mapping can be made from the grammar category to the semantics category. Furthermore, the diagrammatic calculus can also be used in this context.

Positive operators were proposed in [1, 2] as a means of representing word meanings since they have a natural ordering called the Löwner ordering. This ordering states that for two positive operators $A$ and $B$,

$$A \leqslant B \iff B - A \text{ is positive}$$

This ordering can be used to represent hyponymy and lexical entailment. In [1, 25] concrete proposals for building positive operators representing words are given.

The space of positive operators and the properties of the Löwner ordering on this space has been examined in [13, 40]. When the set of positive operators is restricted to those with maximum eigenvalue less than or equal to 1, the ordering has nice properties. We restrict to this set, and use the notation $\mathcal{CP}_1(V)$. When the set of positive operators is restricted to those with eigenvalues exactly 1, we have the set of projection operators onto subspaces of $V$, and the Löwner ordering corresponds to subspace inclusion on projection operators.

The Löwner ordering is crisp: either the relation obtains or it doesn't. However, when considering natural language, we are also interested in graded notions of hyponymy and entailment. For example, although we may consider *dog* to be highly indicative of *pet*, not every dog is a pet, and so we want some kind of graded ordering. On the other hand, we would expect *dog* to be a full hyponym of *mammal*.

[1] introduce a graded notion of hyponymy based on the relative entropy of two operators. [2] use a graded notion of hyponymy that is based on expanding the hypernym (the broader term) to include the hyponym. [25] extends this idea to include a wider range of gradings.

Specifically, suppose we are comparing two positive operators $A$ and $B$. If $A \leqslant B$ crisply, then $B = A + D$ for some positive operator $D$. However, if this is not the case, then we can consider an error term $E$ so that now

$$A + D = B + E$$

Then we have that $B - A = D - E$, i.e. that there is a wholly positive and a wholly negative component to the difference $B - A$. In [2] the authors render the error term $E$ as being of the form $(1-k)A$, where $k \in [0, 1)$. Then the value $k$ is the strength of the hyponymy relation between $A$ and $B$. The drawback of this approach is that the span of $A$ must be included within the span of $B$. [25] proposes two alternative gradings based on the error term that do not suffer from this drawback:

$$k_{BA}(A, B) = \frac{Tr(D - E)}{Tr(D + E)} \qquad (4)$$

$$k_E(A, B) = 1 - \frac{||E||}{||A||} \qquad (5)$$

In equation (4), in the worst case the positive difference term $D$ is 0, and then $k_{BA} = -1$. In the best case $E = 0$ and then $k_{BA} = 1$. Furthermore, the worst case is obtained when the operator $B$ is the zero matrix. In equation (5), in the worst case $E = A$, and then $k_E = 0$. In the best case $E = 0$ and then $k_E = 1$. This means that in both cases the best score of 1 is obtained when the crisp Löwner order obtains. We can think of $A$ as being contained in $B$. For $k_E$ the worst case obtains when the spans of the operators of the words are disjoint. So we can think of this as saying that word $A$ is unrelated to word $B$. However, for $k_{BA}$, the score becomes smaller as $Tr(B)$ becomes smaller. Therefore, even if the spans of the operators are disjoint, the lowest value is obtained only when $Tr(B) = 0$. This means that when measuring whether, say, *dog* is a hyponym of *mathematics*, $k_E$ will (hopefully) give back a value of 0. However, $k_{BA}$ will not result in the value $-1$, because (hopefully) $Tr(\llbracket mathematics \rrbracket) \neq 0$. It would therefore seem that $k_E$ is better for representing graded hyponymy. However, whilst this is true at the single word level [25, 24], we will see that $k_{BA}$ works better when words are composed into phrases and sentences.

## 2.3 Building positive operators for words

In [2], a broader term such as *mammal* is viewed as a weighted sum over projectors describing instances of mammals. For example:

$$\llbracket mammal \rrbracket = p_d \left| dog \right\rangle \left\langle dog \right| + p_c \left| cat \right\rangle \left\langle cat \right| + p_w \left| whale \right\rangle \left\langle whale \right| + ...$$
$$\text{where} \quad \forall i . p_i \geq 0 \text{ (and some kind of normalisation may be applied)}$$

[25] propose a means of building positive operators for words using distributional word vectors and information about hyponymy relations from resources such as WordNet [28], as follows. In general, the meaning of a word $w$ is considered to be given by a collection of unit vectors $\{|w_i\rangle \in W\}_i$, where each $|w_i\rangle$ represents an instance of the concept expressed by the word. Then the operator:

$$\llbracket w \rrbracket = \sum_i p_i \left| w_i \right\rangle \left\langle w_i \right| \in W \otimes W \tag{6}$$

represents the word $w$. The $p_i$ are weightings derived from the text, and there are various choices about what these should be.

We build representations of words as positive operators in the following manner. Suppose we have a dictionary of word vectors $\{v_i : |v_i\rangle \in W\}_i$ derived from a corpus using standard distributional or embedding techniques, for example GloVe, [31], fastText [6], or weighted co-occurrence vectors. To build a representation of a word, we obtain a set of hyponyms that are instances of that word. In this paper, we use WordNet [28], a human-curated database of word relationships including hyponym-hypernym pairs. The WordNet hyponymy relationship is naturally arranged as a directed graph with a root (it is not quite a tree). For the noun subset of the database, the root is the most general noun *entity*, and the leaves are specific nouns. For example, under the word *rocket* there are (inter alia): *test_instrument_vehicle*, *Stinger*, *takeoff_booster*, *arugula*. Notice that here we have different meanings of the word *rocket*, one as a projectile and one as a vegetable. There are also less supervised ways of obtaining these relationships using patterns derived from text, see [15, 35] for examples.

To build a positive operator for a word $w$, we go through the WordNet hierarchy and collect all hyponyms $w_i$ of $w$ at all levels. We then form $[\![w]\!]$ as in equation (6), with $p_i = 1$ for all $i$.

When we build these operators, between 1/3 and 1/2 of the hyponyms listed in WordNet are available in the precomputed GloVe vectors that we used[1], since many entries in WordNet are multi-word expressions, which are not included in the set of vectors we used. We therefore miss a large proportion of the information included in WordNet.

## 2.4 Normalization

An important parameter choice is the type of normalization to use. In [2] two choices are discussed: normalizing operators to trace 1, or normalizing operators to have maximum eigenvalue less than or equal to 1. The properties of these two normalization strategies are thoroughly analyzed in [41]. If operators are normalized to trace 1, then the crisp Löwner ordering becomes trivial: no two operators stand in the relation $A \leqslant B$. If operators are normalized to have maximum eigenvalue 1, then the Löwner ordering has particularly nice properties. In the current paper, we will need to normalize operators so that their maximum eigenvalue is less than or equal to 1, as this will allow us to apply our proposed negation operator.

## 2.5 Composing positive operators

Building positive operators as proposed gives us representations for individual words. However, the representations are all states in one object of **CPM**(**FVect**), whereas for verbs, adjectives, and so on, we need morphisms in **CPM**(**FVect**). In order to obtain these, we use an approach outlined in [20]. Firstly, we consider the spaces for noun and sentence to be the same, so now our pregroup types $n$ and $s$ both map to the same space $W$. To represent adjectives and verbs, representations of type $W \otimes W$ or $W \otimes W \otimes W$ are needed. In order to encode our representations in $W \otimes W$, we need to use the word representations we have built to define suitable morphisms in **CPM**(**FVect**). [20] use the notion of a *Frobenius algebra*. Working in **FVect**, a Frobenius algebra over a finite-dimensional vector space with bases $\{|i\rangle\}_i$ is given by

$$\Delta :: |i\rangle \mapsto |i\rangle \otimes |i\rangle \qquad \iota :: |i\rangle \mapsto 1 \qquad \mu :: |i\rangle \otimes |i\rangle \mapsto |i\rangle \qquad \xi :: 1 \mapsto |i\rangle$$

In the graphical calculus, these are given by:

---

[1] available from https://nlp.stanford.edu/projects/glove/

$$\Delta : \quad \text{⋏} \qquad \iota : \quad | \qquad \mu : \quad \text{Y} \qquad \xi : \quad |$$

A vector $|v\rangle \in W$ can be lifted to a higher-order representation in $W \otimes W$ by applying the map $\Delta$. In **FVect**, this higher-order representation takes the vector $|v\rangle$ and embeds it along the diagonal of a matrix in $W \otimes W$. So, for example, given a vector representation of an intransitive verb $|run\rangle \in W$, we can lift that representation to a matrix in $W \otimes W$ by embedding it into the diagonal of a matrix. The Frobenius algebra interacts with the type reduction morphism $\epsilon_N$ in such a way that the result of lifting a verb and then composing with a noun is to apply the $\mu$ multiplication to the tensor product of the noun and the verb vectors, i.e.

$$(\epsilon_N \otimes 1_N) \circ (1_N \otimes \Delta_N)(|noun\rangle \otimes |verb\rangle) = \mu(|noun\rangle \otimes |verb\rangle)$$

Diagrammatically,

In **FVect** the multiplication $\mu$ implements pointwise multiplication of the two vectors. In **CPM(FVect)** we have access to the same algebra, and the multiplication $\mu$ operates similarly - namely, given two positive operators $A$ and $B$, $\mu(A \otimes B)$ implements pointwise multiplication of the two operators. We call this operator **Mult** or $\odot$. Whilst simple and theoretically motivated, this operation is not desirable for some linguistic purposes as it is commutative, so that *dog bites man* gets the same representation as *man bites dog*. It may be suitable for some other combinations, such as conjunctions (but see [45] for extensive work in this area).

In [9, 24], two other multiplications are proposed for combining positive operators. One, which we call **BMult** or $*_B$, was originally proposed in [22, 23] as a quantum Bayesian operation. This takes two operators $A$ and $B$ and returns the non-commutative and non-associative product $B^{\frac{1}{2}} A B^{\frac{1}{2}}$. In [11], the authors show that this operation is also related to a Frobenius algebra, with the caveat that the algebra corresponds to a basis for $W$ that diagonalises $B$. Specifically, consider $B = \sum_i b_i |i\rangle \langle i|$ and $A = \sum_{jk} a_{jk} |j\rangle \langle k|$, and define $C := \sqrt{b} \otimes \sqrt{b}$ where

$\sqrt{b} := diag(B^{\frac{1}{2}})$. Then:

$$B^{\frac{1}{2}} A B^{\frac{1}{2}} = \sum_i \sqrt{b_i} |i\rangle \langle i| \circ \sum_{jk} a_{jk} |j\rangle \langle k| \circ \sum_l \sqrt{b_l} |l\rangle \langle l| \tag{7}$$

$$= \sum_i \sqrt{b_i} |i\rangle \langle i| \circ \sum_{jk} a_{jk} \sqrt{b_k} |j\rangle \langle k| \tag{8}$$

$$= \sum_{jk} a_{jk} \sqrt{b_j} \sqrt{b_k} |j\rangle \langle k| = \mu(A \otimes C) \tag{9}$$

The second composition function, which we call **KMult**, or $*_K$, is to form a completely positive map from a positive matrix $B$ by decomposing $B$ into a weighted sum of orthogonal projectors $B = \sum_i p_i P_i$, and then forming the map

$$\mathcal{B}(A) = \sum_i p_i P_i \circ A \circ P_i$$

If we again consider a basis that diagonalises $B$, this operation then corresponds to the Frobenius multiplication $\mu(A \otimes B)$ in that basis. To see this, consider

$$B = \sum_i b_i |i\rangle \langle i|, \quad A = \sum_{jk} a_{jk} |j\rangle \langle k|$$

Then

$$\mathcal{B}(A) = \sum_i b_i |i\rangle \langle i| \circ \left( \sum_{jk} a_{jk} |j\rangle \langle k| \right) \circ |i\rangle \langle i| \tag{10}$$

$$= \sum_i b_i |i\rangle \langle i| \circ \sum_j a_{ji} |j\rangle \langle i| \tag{11}$$

$$= \sum_i b_i a_{ii} |i\rangle \langle i| = \mu(A \otimes B) \tag{12}$$

We therefore have three ways of combining positive operators. Moreover, each of these combination methods preserves the property that the eigenvalues must be less than or equal to 1. For the operations **Mult** and **KMult**, the spectral radius is submultiplicative with respect to the Hadamard (pointwise) product of two positive semidefinite matrices [18], implying that the maximum eigenvalue of $A \odot B$ is bounded by 1. For the case of **BMult**, note that the product $B^{\frac{1}{2}} A B^{\frac{1}{2}}$ is similar to $AB$ and hence has the same eigenvalues. Then the maximum eigenvalue of the product $AB$ is bounded by the product of the maximum eigenvalues of $A$ and of $B$ [5], again implying that the maximum eigenvalue of $AB$ is bounded by 1.

To apply these multiplications linguistically, choices must be made about the order in which they are applied, since neither **BMult** nor **KMult** are associative. In particular for transitive verbs there are a number of different choices, and some of these are discussed in [25]. For now, we limit to simple intransitive sentences, of the form *noun verb*.

The operators we outlined above are summarised below.

$$\text{Mult: } [\![noun\ verb]\!] = [\![noun]\!] \odot [\![verb]\!] \tag{13}$$

$$\text{BMult: } [\![noun\ verb]\!] = [\![noun]\!] *_B [\![verb]\!] = [\![verb]\!]^{\frac{1}{2}} [\![noun]\!] [\![verb]\!]^{\frac{1}{2}} \tag{14}$$

$$\text{KMult: } [\![noun\ verb]\!] = [\![noun]\!] *_K [\![verb]\!] = \sum_i p_i P_i [\![noun]\!] P_i \tag{15}$$

where in KMult $[\![verb]\!] = \sum_i p_i P_i$.

## 3 Modelling negation in $\mathcal{CP}_1(V)$

So far, we have shown how to build positive operators from a corpus of text, together with information about hyponymy relations. We have also shown how to lift the simple operators thus described to the maps required for functional words such as verbs and adjectives. We now describe how to model negation.

As discussed, one approach to modelling negation is to map a vector to the subspace orthogonal to it. We can incorporate this in our model very easily, since in the case of projectors, this is equivalent to subtracting the associated matrix from the identity matrix. Consider a vector $|dog\rangle$ that we have learnt in a distributional manner from a corpus. We can lift this representation to a positive operator by forming the projector $|dog\rangle\langle dog|$, which forms a one-dimension subspace of the vector space W. We can then form an operator

$$[\![not\ dog]\!] = \mathbb{I} - |dog\rangle\langle dog|$$

which encompasses the $n-1$-dimensional subspace orthogonal to the projector $|dog\rangle\langle dog|$. In the general case, we define

$$[\![not\ w]\!] := \mathbb{I} - [\![w]\!] \tag{16}$$

When we restrict to the subset $\mathcal{CP}_1(W)$ over a vector space $W$, this operation preserves positivity of the operator and also maps operators into the set $\mathcal{CP}_1(W)$.

What does this mean in application to a sentence? If we take a sentence *dogs dance* then the standard ways of adding a negation operator to this sentence would be to go to either: *dogs don't dance* or *no dogs dance*. In the second case, we change the quantification of the sentence: if *dogs dance* his implicitly existentially quantified, then *no dogs dance* is universally quantified, and vice versa. However, at present, we do not have a means of representing quantification in our model, and therefore at present we will steer clear of this interpretation. Another, admittedly less natural, interpretation is to say *non-dogs move*, i.e. negate only the word and not the sentence, and this is the interpretation we will use in this paper.

Importantly, the negation operation we have proposed is *not* a morphism of **CPM(FVect)**, and therefore a suitable home needs to be found for it. We do not provide an answer to that in this paper, leaving it for ongoing work. Rather, we look at how this operation interacts with composition, the Löwner ordering, and how it works in implementation.

## 3.1 How *not* interacts with the (graded) Löwner ordering

Consider operators $A$ and $B \in \mathcal{CP}_1(W)$. Under the crisp Löwner ordering, we have

$$A \leqslant B \iff B = A + D \tag{17}$$
$$\iff \mathbb{I} - B = \mathbb{I} - (A + D) \tag{18}$$
$$\iff \mathbb{I} - B + D = \mathbb{I} - A \iff \textit{not } B \leqslant \textit{not } A \tag{19}$$

Considering an error term $E$, we use the notation $\leqslant_E$ if $B + E = A + D$. With such an error term,

$$A \leqslant_E B \iff B + E = A + D \tag{20}$$
$$\iff \mathbb{I} - (B + E) = \mathbb{I} - (A + D) \tag{21}$$
$$\iff \mathbb{I} - B + D = \mathbb{I} - A + E \iff \textit{not } B \leqslant_E \textit{not } A \tag{22}$$

Depending on the grading we use, the strength of the hyponymy relation will be affected. Using the $k_{BA}$ grading (equation (4)) we have that *not B* is a hyponym of *not A* with strength

$$k_{BA}(\textit{not } B, \textit{not } A) = \frac{Tr(D-E)}{Tr(D+E)} = k_{BA}(A, B)$$

Using $k_E$ (equation (5)), we have:

$$k_E(\textit{not } B, \textit{not } A) = 1 - \frac{||E||}{||\textit{not } B||} \neq k_E(A, B)$$

783

## 3.2 How *not* interacts with composition

We focus here just on the case of intransitive sentences composed of a subject and a verb. When we negate the noun we obtain the following expressions:

$$[\![not\ noun]\!] \odot [\![verb]\!] = (\mathbb{I} - [\![noun]\!]) \odot [\![verb]\!] \tag{23}$$
$$= diag([\![verb]\!]) - [\![noun]\!] \odot [\![verb]\!] \tag{24}$$
$$[\![not\ noun]\!] *_B [\![verb]\!] = (\mathbb{I} - [\![noun]\!]) *_B [\![verb]\!] \tag{25}$$
$$= [\![verb]\!]^{\frac{1}{2}} [\![verb]\!]^{\frac{1}{2}} - [\![verb]\!]^{\frac{1}{2}} [\![noun]\!] [\![verb]\!]^{\frac{1}{2}} \tag{26}$$
$$= [\![verb]\!] - [\![noun]\!] *_B [\![verb]\!] \tag{27}$$
$$[\![not\ noun]\!] *_K [\![verb]\!] = (\mathbb{I} - [\![noun]\!]) *_K [\![verb]\!] \tag{28}$$
$$= \sum_i p_i P_i P_i - \sum_i p_i P_i [\![noun]\!] P_i \tag{29}$$
$$= [\![verb]\!] - [\![noun]\!] *_K [\![verb]\!] \tag{30}$$

Particularly in the case of $*_B$ and $*_K$, these feel like fairly natural interpretations of a sentence with a negated noun. We take the meaning of the verb as a whole, and then subtract out the part of the verb that is applied to the noun.

We can also look at the behaviour of the operators when the verb is negated.

$$[\![noun]\!] \odot [\![not\ verb]\!] = [\![noun]\!] \odot (\mathbb{I} - [\![verb]\!]) \tag{31}$$
$$= diag([\![noun]\!]) - [\![noun]\!] \odot [\![verb]\!] \tag{32}$$

For $*_B$ and $*_K$ we make the assumption that we use a basis in which $[\![verb]\!]$ is diagonal so that $[\![verb]\!] = \sum_i p_i P_i$. Considering the expression in equation (9) we obtain:

$$[\![noun]\!] *_B [\![not\ verb]\!] = ([\![noun]\!]) *_B (\mathbb{I} - [\![verb]\!]) \tag{33}$$
$$= \sum_{jk} a_{jk} \sqrt{1 - p_j} \sqrt{1 - p_k} |j\rangle \langle k| . \tag{34}$$

This expression does not relate very well to the semantics of the particular words. We do obtain something better with $*_K$:

$$[\![noun]\!] *_K [\![not\ verb]\!] = ([\![noun]\!]) *_K (\mathbb{I} - [\![verb]\!]) \tag{35}$$
$$= \sum_i (1 - p_i) P_i [\![noun]\!] P_i \tag{36}$$
$$= \sum_i P_i [\![noun]\!] P_i - \sum_i p_i P_i [\![noun]\!] P_i \tag{37}$$
$$= diag([\![noun]\!]) - [\![noun]\!] *_K [\![verb]\!] . \tag{38}$$

The operation $*_B$ does not have a particularly illuminating representation when the verb is negated, but in the case of $\odot$ and $*_K$, these are again fairly natural interpretations of a sentence with a negated verb.

## 4 Demonstrations

We give a demonstration on a small dataset that this rendering of negation works well together with the composition operators proposed. In particular, we will see that our combination operators can beat baselines that examine just the noun or the verb in the sentence. This is an important baseline since the construction of the dataset is such that entailment does follow from comparing either the nouns or the verbs. Our combination operators do not in general beat an average of two operators, however, they do in some cases.

### 4.1 Datasets

We build a set of datasets based on the intransitive sentence dataset introduced in [37]. The dataset consists of paired sentences consisting of a subject and a verb. In half the cases the first sentence entails the second, and in the other half of cases, the order of the sentences is reversed. For example, we have:

summer finish, season end, T

season end, summer finish, F

The first sentence is marked as entailing, whereas the second is marked as not entailing. The dataset is created by selecting nouns and verbs from WordNet. In the case of the sentence marked T, the first noun is selected as a hyponym of the second noun, and the first verb is selected as a hyponym of the second verb.

For these sentences to be thought of as entailing, we must view them as being implicitly existentially quantified. For example, if we took the pair of sentences

gazelles sprint, mammals run

we can clearly see that the first sentence does not entail the second if we assume a universal quantification - there could easily be, and there are, non-gazelle mammals that don't run. However, if we take an existential quantification, then the fact that there is some gazelle that sprints means that there must be some mammal (the gazelle) who runs (as sprinting is a kind of running).

Bearing in mind that the sentences are existentially quantified, we create three further datasets that include negation. We apply negation only at the word level and not at the sentence level, as this retains the existentially quantified nature of the sentences. Consider an entailing sentence pair such as:

$$dogs\ run \models mammals\ move$$

We include negation in two places: either the noun can be negated, giving us *non-dogs* and *non-mammals*, or else the verbs can be negated, giving us *do not run* and *do not move*.

From *dogs run* $\models$ *mammals move* we then get three more pairs of entailing sentences:

$$some\ dogs\ run \models some\ mammals\ move \tag{39}$$
$$some\ non\text{-}mammals\ run \models some\ non\text{-}dogs\ move \tag{40}$$
$$some\ dogs\ do\ not\ move \models some\ mammals\ do\ not\ run \tag{41}$$
$$some\ non\text{-}mammals\ do\ not\ move \models some\ non\text{-}dogs\ do\ not\ run \tag{42}$$

To model these, we render the negation of the verb as directly acting on the verb. Another choice would be for the negation to act on the whole sentence, rendering *dogs don't move* as *not(dogs move)*, but this would mean that we now consider the sentence universally quantified. Working out how to include a full account of quantification is an area of further work.

To model these sentences, we therefore calculate, respectively:

$$[\![dogs]\!] * [\![run]\!] \leqslant_k [\![mammals]\!] * [\![move]\!] \tag{43}$$
$$(\mathbb{I} - [\![mammals]\!]) * [\![run]\!] \leqslant_k (\mathbb{I} - [\![dogs]\!]) * [\![move]\!] \tag{44}$$
$$[\![dogs]\!] * (\mathbb{I} - [\![move]\!]) \leqslant_k [\![mammals]\!] * (\mathbb{I} - [\![run]\!]) \tag{45}$$
$$(\mathbb{I} - [\![mammals]\!]) * (\mathbb{I} - [\![move]\!]) \leqslant_k (\mathbb{I} - [\![dogs]\!]) * (\mathbb{I} - [\![run]\!]) \tag{46}$$

where $\leqslant_k \in \{k_{BA}, k_E\}$ is one of the graded hyponymy measures and $* \in \{\odot, *_B, *_K\}$ is one of the compositional operators.

## 4.2 Construction and composition of positive operators

We follow the construction methods outlined in [25] and summarised in this paper in section 2.3. In order to construct the basic positive operators, we use hyponyms from WordNet [28], and 50 or 300 dimensional GloVe vectors. The operators produced are normalised to have maximum eigenvalue equal to 1.

To compose positive operators, we use the three composition functions **Mult**, **BMult**, **KMult** discussed in section 2.5. We compare these with three baselines:

| Model | noun-verb | ¬noun-verb | noun-¬verb | ¬noun-¬verb |
|---|---|---|---|---|
| KS2016 best | 0.84 | - | - | - |
| Verb only | 0.866 | 0.867 | 0.865 | 0.867 |
| Noun only | 0.926 | 0.921 | 0.925 | 0.923 |
| Average | 0.947$^+$ | **0.946$^+$** | **0.948$^+$** | 0.946$^+$ |
| Mult | **0.960$^{*+}$** | 0.874 | 0.931$^+$ | **0.950$^+$** |
| BMult | 0.948$^+$ | 0.892 | 0.928 | 0.947$^+$ |
| BMult switched | 0.949$^+$ | 0.896 | 0.916 | 0.944$^+$ |
| KMult | 0.950$^+$ | 0.875 | 0.925 | 0.948$^+$ |
| KMult switched | 0.950$^+$ | 0.874 | 0.920 | 0.948$^+$ |

Table 1: Area under ROC curve on the negation datasets, using $k_{BA}$, WordNet hyponyms, and 300 dimensional GloVe vectors. Figures reported are the average of the 100 values of the test statistic. * indicates significantly better than the Average baselline. $^+$ indicates significantly better than the noun-only baseline.

the average of two operators, a noun-only baseline, and a verb-only baseline. Due to the construction of the datasets, we see that in fact the verb-only and noun-only baselines are fairly strong, since as long as the construction of the individual words models the hyponymy relations well then a verb-only or noun-only model will be able to perform well on these datasets. Note that taking the average of the two operators preserves the criterion of the maximum eigenvalue being less than or equal to 1 by Weyl's inequalities [42].

**Metrics and significance measures** Since the entailment measures we use give back a grading, whereas we require a binary response, we calculate area under ROC curve (AUC). The AUC calculates the true positive rate vs. the false positive rate for different cutoff levels of the graded measure. The maximum that can be attained is 1.

To measure the significance of our results, we use bootstrapping [14] to calculate 100 values of the test statistic (AUC) drawn from the distribution implied by the data. We compare between models using a paired t-test and apply the Bonferroni correction to compensate for multiple model comparisons.

## 5 Results

We can see that across the board (tables 1, 2, 3, 4), the $k_{BA}$ measure performs more strongly than the $k_E$ measure. The difference in performance is likely to be because the $k_{BA}$ measure is very symmetric, and the dataset is also, meaning that not only

| Model | noun-verb | ¬noun-verb | noun-¬verb | ¬noun-¬verb |
|---|---|---|---|---|
| KS2016 best | 0.84 | - | - | - |
| Verb only | 0.635 | 0.637 | 0.636 | 0.634 |
| Noun only | 0.686 | 0.643 | 0.684 | 0.635 |
| Average | 0.727 | 0.778$^+$ | 0.777$^+$ | 0.782$^+$ |
| Mult | **0.883**$^{*+}$ | **0.885**$^{*+}$ | 0.899$^{*+}$ | **0.952**$^{*+}$ |
| BMult | 0.792$^{*+}$ | 0.678$^+$ | 0.725$^+$ | 0.719$^+$ |
| BMult switched | 0.786$^{*+}$ | 0.693$^+$ | 0.715$^+$ | 0.718$^+$ |
| KMult | 0.873$^{*+}$ | 0.725$^+$ | **0.900**$^{*+}$ | 0.732$^+$ |
| KMult switched | 0.839$^{*+}$ | 0.879$^{*+}$ | 0.732$^+$ | 0.666$^+$ |

Table 2: Area under ROC curve on the negation datasets, using $k_E$, WordNet hyponyms, and 300 dimensional GloVe vectors. Figures reported are the average of the 100 values of the test statistic. * indicates significantly better than the Average baselline. $^+$ indicates significantly better than the noun-only baseline.

| Model | noun-verb | ¬noun-verb | noun-¬verb | ¬noun-¬verb |
|---|---|---|---|---|
| KS2016 best | 0.84 | - | - | - |
| Verb only | 0.787 | 0.787 | 0.786 | 0.785 |
| Noun only | 0.907 | 0.906 | 0.903 | 0.904 |
| Average | 0.929$^+$ | **0.925**$^+$ | 0.929$^+$ | 0.930$^+$ |
| Mult | **0.942**$^{*+}$ | 0.836 | 0.915$^+$ | 0.925$^+$ |
| BMult | 0.917$^+$ | 0.861 | 0.914$^+$ | 0.920$^+$ |
| BMult switched | 0.918$^+$ | 0.859 | 0.912$^+$ | 0.922$^+$ |
| KMult | 0.929$^+$ | 0.829 | 0.910$^+$ | 0.926$^+$ |
| KMult switched | 0.926$^+$ | 0.821 | 0.911$^+$ | 0.930$^+$ |

Table 3: Area under ROC curve on the negation datasets, using $k_{BA}$, WordNet hyponyms, and 50 dimensional GloVe vectors. Figures reported are the average of the 100 values of the test statistic. * indicates significantly better than the Average baselline. $^+$ indicates significantly better than the noun-only baseline.

are there equal numbers of entailing and non-entailing sentences in the dataset, but the non-entailing datasets are the opposite of the entailing datasets. Enhancing the datasets with some random pairings would likely degrade the performance of the $k_{BA}$ measure. Investigating the differences in performance in a less balanced dataset is an area of further work.

In the case of the $k_{BA}$ measure, increasing the dimensionality of the underlying vector space improved performance across all sentence types. This was not the case for the $k_E$ measure, where for sentences of the type *noun - not verb* and *not noun - verb* performance using the $k_E$ measure improved with lower dimension-

| Model | noun-verb | ¬noun-verb | noun-¬verb | ¬noun-¬verb |
|---|---|---|---|---|
| KS2016 best | 0.84 | - | - | - |
| Verb only | 0.601 | 0.605 | 0.605 | 0.607 |
| Noun only | 0.708 | 0.724 | 0.706 | 0.720 |
| Average | 0.753$^+$ | 0.791$^+$ | 0.783$^+$ | 0.797$^+$ |
| Mult | 0.847$^{*+}$ | **0.845**$^{*+}$ | **0.891**$^{*+}$ | **0.925**$^{*+}$ |
| BMult | 0.751$^+$ | 0.694 | 0.738$^+$ | 0.751$^+$ |
| BMult switched | 0.728$^+$ | 0.707 | 0.727$^+$ | 0.758$^+$ |
| KMult | **0.875**$^{*+}$ | 0.702 | 0.875$^{*+}$ | 0.802$^+$ |
| KMult switched | 0.808$^{*+}$ | 0.791$^{*+}$ | 0.726$^+$ | 0.815$^{*+}$ |

Table 4: Area under ROC curve on the negation datasets, using $k_E$, WordNet hyponyms, and 50 dimensional GloVe vectors. Figures reported are the average of the 100 values of the test statistic. * indicates significantly better than the Average baseline. + indicates significantly better than the noun-only baseline.

ality (tables 2 and 4)

The best results were obtained using the $k_{BA}$ measure and 300-dimensional GloVe vectors. In this set of results (table 1) the Average baseline proves hard to beat, however Mult also performs strongly for sentences with either no word negated or both words negated. For these two classes of sentences, it is also notable that all composition functions enable better performance than the strong non-compositional noun-only baseline. A similar pattern is seen when using 50-dimensional vectors with the $k_{BA}$ measure (table 3), where the benefit of using a compositional operator is also seen for the sentence type *noun - not verb*.

The benefit of using compositional operators is also seen for the $k_E$ measure (tables 2 and 4), where using a compositional operator helps in almost all cases over the (admittedly much worse) non-compositional noun-only baseline.

Across both measures and dimensionalities performance is poor on the sentence type *not noun - verb*. More research is needed to investigate why this is.

## 6 Discussion and Further Work

We have introduced a negation operator for use in the **CPM(FVect)** flavour of DisCoCat. The operators is based on the notion of projection onto the orthogonal subspace, used previously by [44]. The operator works well together with the composition operators Mult, BMult, and KMult discussed in [24, 25, 11], and in many cases perform well on a toy dataset of sentence entailments.

More investigation into the properties of the **BMult** and **KMult** operators is needed. In [11] it was shown that the two operators can be combined together in a double density matrix setting, meaning that the operators can be given a natural home.

Further, testing on larger scale datasets is also needed. Ideally, the kinds of entailment relations we are looking at should be useful for textual entailment and reasoning systems. Expanding the models we currently have to test on realistic datasets is desirable. One move towards this would be use these operators with the Stanford Natural Language Inference (SNLI) dataset [8]. That dataset has pairs of short sentences labelled either as entailing, contradictory, or neutral. The sentences are given a binary parse, determining the order of composition. Example sentences from the dataset where the first entails the second are:

$S1$ :( Children ( ( ( smiling and ) waving ) ( at camera ) ) )
$S2$ :( There ( ( are children ) present ) )

Given positive operators for each word, we would be able to combine them together in the specified order to form sentence representations, and use these to predict the relationship that obtains between sentences.

In order to build operators for a wider range of words, techniques that can generate representations directly from a corpus are needed. At present, data on hyponymy relations from WordNet is used. Work is ongoing to build operators directly from a corpus, with success in learning operator representations that model ambiguity well [27].

Another major unanswered question is where the negation operator should sit theoretically. It cannot be viewed as a morphism in **CPM(FVect)**. Some work in progress is into looking at the set $\mathcal{CP}_1(W)$ as an object of the category **ConvexRel**, introduced in [7]. Then, the negation operator can be viewed as a morphism. This is an area of further work.

# References

[1] Esma Balkır, Mehrnoosh Sadrzadeh, and Bob Coecke. Distributional sentence entailment using density matrices. In *Topics in Theoretical Computer Science*, pages 1–22. Springer, 2016.

[2] Dea Bankova, Bob Coecke, Martha Lewis, and Dan Marsden. Graded hyponymy for compositional distributional semantics. *Journal of Language Modelling*, 6(2):225–260, 2019.

[3] Marco Baroni and Roberto Zamparelli. Nouns are vectors, adjectives are matrices: Representing adjective-noun constructions in semantic space. In *Proceedings of the*

*2010 Conference on Empirical Methods in Natural Language Processing*, pages 1183–1193. Association for Computational Linguistics, 2010.

[4] Islam Beltagy, Stephen Roller, Gemma Boleda, Katrin Erk, and Raymond Mooney. Utexas: Natural language semantics using distributional semantics and probabilistic logic. In *Proceedings of the 8th International Workshop on Semantic Evaluation (SemEval 2014)*, pages 796–801, 2014.

[5] Rajendra Bhatia. *Matrix analysis*, volume 169. Springer Science & Business Media, 2013.

[6] Piotr Bojanowski, Edouard Grave, Armand Joulin, and Tomas Mikolov. Enriching word vectors with subword information. *Transactions of the Association for Computational Linguistics*, 5:135–146, 2017.

[7] Joe Bolt, Bob Coecke, Fabrizio Genovese, Martha Lewis, Dan Marsden, and Robin Piedeleu. Interacting conceptual spaces i: Grammatical composition of concepts. In *Conceptual Spaces: Elaborations and Applications*, pages 151–181. Springer, 2019.

[8] Samuel Bowman, Gabor Angeli, Christopher Potts, and Christopher D Manning. A large annotated corpus for learning natural language inference. In *Proceedings of the 2015 Conference on Empirical Methods in Natural Language Processing*, pages 632–642, 2015.

[9] Bob Coecke. The mathematics of text structure. *arXiv preprint arXiv:1904.03478*, 2019.

[10] Bob Coecke, Edward Grefenstette, and Mehrnoosh Sadrzadeh. Lambek vs. Lambek: Functorial vector space semantics and string diagrams for Lambek calculus. *Annals of Pure and Applied Logic*, 164(11):1079–1100, 2013.

[11] Bob Coecke and Konstantinos Meichanetzidis. Meaning updating of density matrices. *arXiv preprint arXiv:2001.00862*, 2020.

[12] Bob Coecke, Mehrnoosh Sadrzadeh, and Stephen Clark. Mathematical foundations for a compositional distributional model of meaning. *arXiv:1003.4394*, 2010.

[13] Ellie D'Hondt and Prakash Panangaden. Quantum weakest preconditions. *Mathematical Structures in Computer Science*, 16(03):429–451, 2006.

[14] Bradley Efron. Bootstrap methods: another look at the jackknife. In *Breakthroughs in statistics*, pages 569–593. Springer, 1992.

[15] Marti A Hearst. Automatic acquisition of hyponyms from large text corpora. In *Proceedings of the 14th conference on Computational linguistics-Volume 2*, pages 539–545. Association for Computational Linguistics, 1992.

[16] Jules Hedges and Mehrnoosh Sadrzadeh. A generalised quantifier theory of natural language in categorical compositional distributional semantics with bialgebras. *Mathematical Structures in Computer Science*, 29(6):783–809, 2019.

[17] Karl Moritz Hermann, Edward Grefenstette, and Phil Blunsom. "not not bad" is not "bad": A distributional account of negation. *arXiv preprint arXiv:1306.2158*, 2013.

[18] Roger A Horn and Charles R Johnson. *Matrix analysis*. Cambridge University Press, 1985.

[19] Dimitri Kartsaklis. Coordination in categorical compositional distributional semantics. *arXiv preprint arXiv:1606.01515*, 2016.

[20] Dimitri Kartsaklis, Mehrnoosh Sadrzadeh, and Stephen Pulman. A unified sentence space for categorical distributional-compositional semantics: Theory and experiments. In *In Proceedings of COLING: Posters*, pages 549–558, 2012.

[21] Germán Kruszewski, Denis Paperno, Raffaella Bernardi, and Marco Baroni. There is no logical negation here, but there are alternatives: Modeling conversational negation with distributional semantics. *Computational Linguistics*, 42(4):637–660, December 2016.

[22] Matthew S Leifer and David Poulin. Quantum graphical models and belief propagation. *Annals of Physics*, 323(8):1899–1946, 2008.

[23] Matthew S Leifer and Robert W Spekkens. Towards a formulation of quantum theory as a causally neutral theory of bayesian inference. *Physical Review A*, 88(5):052130, 2013.

[24] M. Lewis. Hyponymy in DisCoCat, 2019. Under review.

[25] Martha Lewis. Compositional hyponymy with positive operators. In *Proceedings of the International Conference on Recent Advances in Natural Language Processing (RANLP 2019)*, pages 638–647, Varna, Bulgaria, September 2019. INCOMA Ltd.

[26] Jean Maillard, Stephen Clark, and Edward Grefenstette. A type-driven tensor-based semantics for ccg. In *Proceedings of the EACL 2014 Workshop on Type Theory and Natural Language Semantics (TTNLS)*, pages 46–54, 2014.

[27] Francois Meyer. Lexical Ambiguity with Density Matrices. Master's thesis, University of Amsterdam, 2020.

[28] George A. Miller. Wordnet: A lexical database for english. *Commun. ACM*, 38(11):39–41, November 1995.

[29] Reinhard Muskens and Mehrnoosh Sadrzadeh. Context update for lambdas and vectors. In *International Conference on Logical Aspects of Computational Linguistics*, pages 247–254. Springer, 2016.

[30] Denis Paperno, Marco Baroni, et al. A practical and linguistically-motivated approach to compositional distributional semantics. In *Proceedings of the 52nd Annual Meeting of the Association for Computational Linguistics (Volume 1: Long Papers)*, volume 1, pages 90–99, 2014.

[31] Jeffrey Pennington, Richard Socher, and Christopher D. Manning. Glove: Global vectors for word representation. In *Empirical Methods in Natural Language Processing (EMNLP)*, pages 1532–1543, 2014.

[32] Robin Piedeleu, Dimitri Kartsaklis, Bob Coecke, and Mehrnoosh Sadrzadeh. Open system categorical quantum semantics in natural language processing. *arXiv:1502.00831*, 2015.

[33] Anne Preller and Mehrnoosh Sadrzadeh. Bell states and negative sentences in the distributed model of meaning. *Electronic Notes in Theoretical Computer Science*, 270(2):141–153, 2011.

[34] Laura Rimell, Amandla Mabona, Luana Bulat, and Douwe Kiela. Learning to negate

adjectives with bilinear models. In *Proceedings of the 15th Conference of the European Chapter of the Association for Computational Linguistics: Volume 2, Short Papers*, pages 71–78, 2017.

[35] Stephen Roller, Douwe Kiela, and Maximilian Nickel. Hearst patterns revisited: Automatic hypernym detection from large text corpora. *arXiv preprint arXiv:1806.03191*, 2018.

[36] Mehrnoosh Sadrzadeh, Stephen Clark, and Bob Coecke. The Frobenius anatomy of word meanings I: subject and object relative pronouns. *Journal of Logic and Computation*, page ext044, 2013.

[37] Mehrnoosh Sadrzadeh, Dimitri Kartsaklis, and Esma Balkir. Sentence entailment in compositional distributional semantics. *Ann. Math. Artif. Intell.*, 82(4):189–218, 2018.

[38] Peter Selinger. Dagger compact closed categories and completely positive maps. *Electronic Notes in Theoretical Computer Science*, 170:139–163, 2007.

[39] Peter Selinger. A survey of graphical languages for monoidal categories. In *New structures for physics*, pages 289–355. Springer, 2010.

[40] John van de Wetering. Entailment relations on distributions. *arXiv preprint arXiv:1608.01405*, 2016.

[41] John van de Wetering. Ordering information on distributions. *arXiv preprint arXiv:1701.06924*, 2017.

[42] Hermann Weyl. Das asymptotische verteilungsgesetz der eigenwerte linearer partieller differentialgleichungen (mit einer anwendung auf die theorie der hohlraumstrahlung). *Mathematische Annalen*, 71(4):441–479, 1912.

[43] Dominic Widdows and Trevor Cohen. Reasoning with vectors: A continuous model for fast robust inference. *Logic Journal of the IGPL*, 23(2):141–173, 2015.

[44] Dominic Widdows and Stanley Peters. Word vectors and quantum logic: Experiments with negation and disjunction. *Mathematics of language*, 8(141-154), 2003.

[45] Gijs Wijnholds and Mehrnoosh Sadrzadeh. A type-driven vector semantics for ellipsis with anaphora using lambek calculus with limited contraction. *Journal of Logic, Language and Information*, 28(2):331–358, 2019.

# DENSITY MATRICES WITH METRIC FOR DERIVATIONAL AMBIGUITY

ADRIANA D. CORREIA [*]
*Institute for Theoretical Physics and Center for Complex Systems Studies, Utrecht University, P.O. Box 80.089, 3508 TB, Utrecht, The Netherlands.*

MICHAEL MOORTGAT
*Utrecht Institute of Linguistics OTS and Center for Complex Systems Studies, Utrecht University, 3512 JK, Utrecht, The Netherlands*

HENK T. C. STOOF
*Institute for Theoretical Physics and Center for Complex Systems Studies, Utrecht University, P.O. Box 80.089, 3508 TB, Utrecht, The Netherlands.*

## Abstract

Recent work on vector-based compositional natural language semantics has proposed the use of density matrices to model lexical ambiguity and (graded) entailment. Ambiguous word meanings, in this work, are represented as mixed states, and the compositional interpretation of phrases out of their constituent parts takes the form of a strongly monoidal functor sending the derivational morphisms of a pregroup syntax to linear maps in FdHilb.

Our aims in this paper are threefold. Firstly, we replace the pregroup front end by a Lambek categorial grammar with directional implications expressing a word's selectional requirements. By the Curry-Howard correspondence, the derivations of the grammar's type logic are associated with terms of the (ordered) linear lambda calculus; these terms can be read as programs for compositional meaning assembly with density matrices as the target semantic spaces. Secondly, we extend on the existing literature and introduce a symmetric, non-degenerate bilinear form called a "metric" that defines a canonical isomorphism between a vector space and its dual, allowing us to keep a distinction between left and right implication. Thirdly, we use this metric to define density matrix spaces in a directional form, modeling the ubiquitous derivational ambiguity of natural language syntax, and show how this allows an integrated treatment of lexical and derivational forms of ambiguity controlled at the level of the interpretation.

---

[*]Corresponding author: a.duartecorreia@uu.nl

## 1 Introduction

Semantic representations of language using vector spaces are an increasingly popular approach to automate natural language processing, with early comprehensive accounts given in [4, 16]. This idea has found several implementations, both theoretically and computationally. On the theoretical side, the principle of compositionality [12] states that the meaning of a complex expression can be computed from the meaning of its simpler building blocks and the rules used to assemble them. On the computational side, the distributional hypothesis [11] asserts that a meaning of a word is adequately represented by looking at what words most often appear next to it. Joining these two approaches, a distributional compositional categorical (DisCoCat) model of meaning has been proposed [5], mapping the pregroup algebra of syntax to vectors spaces with tensor operations, by functorialy relating the properties of the categories that describe those structures, allowing one to interpret compositionality in a grammar-driven manner using data-extracted representations of words that are in principle agnostic to grammar. This method has been shown to give good results when used to compare meanings of complex expressions and with human judgements [10]. Developments in the computation of these vectors that use machine learning algorithms [15] provide representations of words that start deviating from the count-based models. However, each model still provides a singular vector embedding for each word, which allows the DisCoCat model to be applied with some positive results [30].

The principal limitation of these embeddings, designated *static* embeddings, is that it provides the same word representation independently of context. This hides polysemy, or even subtler gradations in meaning. Using the DisCoCat framework, this issue has been tackled using density matrices to describe lexical ambiguity [22, 23], and using the same framework also sentence entailment [24] and graded hyponymy [1], since the use of matrices allows the inclusion of correlations between context words. From the computational side, the most recent computational language models [7, 21] present contextual embeddings of words as an intrinsic feature. In this paper we aim at reconciling the compositional distributional model and these developments by presenting density matrices as the fundamental representations of words, thus leveraging previous results, and by introducing a refined notion of tensor contraction that can be applied even if we do not assume that we are working with static embeddings coming from the data, thus additionally presenting the possibility of eliminating the distinction between context and target words, because all words can be equally represented with respect to one another. To achieve this, we build the components of the density matrices as covariant or contravariant by introducing a metric that relates them, extending to the interpretation space the notion of directionality of word application, as a direct image of the directional Lambek calculus. After that, we attach permutation operations that act on either type of components to describe derivational ambiguity in a way that keeps multiple readings represented in formally independent vector spaces, thus opening up the possibility of integration between lexical and syntactic ambiguity.

Section 2 introduces our syntactic engine, the Lambek calculus $(\mathbf{N})\mathbf{L}_{/,\backslash}$, together with the Curry-Howard correspondence that associates syntactic derivations with programs of the ordered lambda calculus $\lambda_{/,\backslash}$. Section 3 motivates the use of a more refined notion of inner product and introduces the concept of a tensor and tensor contraction as a basis independent application of a dual vector to a vector, and introduces a metric as the mechanism to go from vectors extracted from the data to the dual vectors necessary to perform tensor contraction. Section 4 gives some background on density matrices, and on ways of capturing the directionality of our syntactic type logic in these semantic spaces using the previously described metric. Section 5 then turns to the compositional interpretation of the $\lambda_{/,\backslash}$ programs associated with $(\mathbf{N})\mathbf{L}_{/,\backslash}$ derivations. Section 6 shows how the directional density matrix framework can be used to capture simple forms of derivational ambiguity.

## 2  From proofs to programs

With his [13, 14] papers, Jim Lambek initiated the 'parsing as deduction' method in computational linguistics: words are assigned formulas of a type logic designed to reason about grammatical composition; the judgement whether a phrase is well-formed is the outcome of a process of deduction in that type logic. Lambek's original work was on a calculus of *syntactic* types, which he presented in two versions. With $\mathbf{L}_{/,\backslash}$ we refer to the simply typed (implicational) fragment of Lambek's [13] associative syntactic calculus, which assigns types to *strings*; $\mathbf{NL}_{/,\backslash}$ is the non-associative version of [14], where types are assigned to *phrases* (bracketed strings).[1]

Van Benthem [27] added semantics to the equation with his work on **LP**, a commutative version of the Lambek calculus, which in retrospect turns out to be a precursor of (multiplicative intuitionistic) linear logic. **LP** is a calculus of *semantic* types. Under the Curry-Howard 'proofs-as-programs' approach, derivations in **LP** are in 1-to-1 correspondence with terms of the (linear) lambda calculus; these terms

---

[1] Neither of these calculi by itself is satisfactory for modelling natural language syntax. To handle the well-documented problems of over/undergeneration of $(\mathbf{N})\mathbf{L}_{/,\backslash}$ in a principled way, the logics can be extended with modalities that allow for controlled forms of reordering and/or restructuring. We address these extensions in [6].

> Terms: $t, u ::= x \mid \lambda^r x.t \mid \lambda^l x.t \mid t \triangleleft u \mid u \triangleright t$
>
> Typing rules:
>
> $$\overline{x : A \vdash x : A} \, Ax$$
>
> $$\frac{\Gamma, x : A \vdash t : B}{\Gamma \vdash \lambda^r x.t : B/A} \, I/ \qquad \frac{x : A, \Gamma \vdash t : B}{\Gamma \vdash \lambda^l x.t : A\backslash B} \, I\backslash$$
>
> $$\frac{\Gamma \vdash t : B/A \quad \Delta \vdash u : A}{\Gamma, \Delta \vdash t \triangleleft u : B} \, E/ \qquad \frac{\Gamma \vdash u : A \quad \Delta \vdash t : A\backslash B}{\Gamma, \Delta \vdash u \triangleright t : B} \, E\backslash$$

Figure 1: Proofs as programs for $(\mathbf{N})\mathbf{L}_{/,\backslash}$.

can be seen as *programs* for compositional meaning assembly. To establish the connection between syntax and semantics, the Lambek-Van Benthem framework relies on a homomorphism sending types and proofs of the syntactic calculus to their semantic counterparts.

In this paper, rather than defining semantic interpretation on a commutative type logic such as **LP**, we want to keep the distinction between the left and right implications $\backslash, /$ of the syntactic calculus in the vector-based semantics we aim for. To achieve this, our programs for meaning composition use the language of Wansing's [29] *directional* lambda calculus $\lambda_{/,\backslash}$. Wansing's overall aim is to study how the derivations of a family of substructural logics can be encoded by typed lambda terms. Formulas, in the substructural setting, are seen as information pieces, and the proofs manipulating these formulas as information processing mechanisms, subject to certain conditions that reflect the presence or absence of structural rules. The terms of $\lambda_{/,\backslash}$ faithfully encode proofs of $(\mathbf{N})\mathbf{L}_{/,\backslash}$; information pieces, in these logics, cannot be copied or deleted (absence of Contraction and Weakening), and information processing is sensitive to the sequential order in which the information pieces are presented (absence of Permutation).

We present the rules of $(\mathbf{N})\mathbf{L}_{/,\backslash}$ with the associated terms of $\lambda_{/,\backslash}$ in Fig 1. The presentation is in the sequent-style natural deduction format. The formula language has atomic types (say *s*, *np*, *n* for sentences, noun phrases, common nouns) for complete expressions and implicational types $A\backslash B$, $B/A$ for incomplete expressions, selecting an $A$ argument to the left (resp. right) to form a $B$.

Ignoring the term labeling for a moment, judgments are of the form $\Gamma \vdash A$, where the antecedent $\Gamma$ is a non-empty list (for **L**) or bracketed list (**NL**) of formulas, and the succedent a single formula $A$. For each of the type-forming operations, there is an Introduction rule, and an Elimination rule.

Turning to the Curry-Howard encoding of $\mathbf{NL}_{/,\backslash}$ proofs, we introduce a language of directional lambda terms, with variables as atomic expressions, left and right $\lambda$ abstraction, and left and right application. The inference rules now become *typing* rules for these terms, with judgments of the form

$$x_1 : A_1, \ldots, x_n : A_n \vdash t : B. \tag{1}$$

The antecedent is a typing environment providing type declarations for the variables $x_i$; a proof constructs a program $t$ of type $B$ out of these variables. In the absence of Contraction, Weakening and Permutation structural rules, the program $t$ contains $x_1, \ldots, x_n$ as free variables exactly once, and in that order. Intuitively, one can see a term-labelled proof as an algorithm to compute a meaning $t$ of type $B$ with parameters $x_i$ of type $A_i$. In parsing a particular phrase, one substitutes the meaning of the constants (i.e. words) that make it up for the parameters of this algorithm.

## 3 Directionality in interpretation

In order to introduce the directionality of the syntactic calculus in the semantic calculus, we expand on the existing literature that uses **FdVect** as the interpretation category by calling attention to the implied inner product. We introduce a more abstract notion of tensor, tensor contraction and the need to introduce explicitly the existence of a metric, coming from the literature of general relativity, following the treatment in [28].[2] Formally, a metric is a function that assigns a distance between two elements of a set, but if applied to the elements of a set that is closed under addition and scalar multiplication, that is, the elements of a vector space, it becomes an inner product. Since we will be looking at vector spaces, we use the terms metric and inner product interchangeably.

To motivate the need for a more careful treatment regarding the inner product, lets look at a very simple yet illustrative example. Suppose that a certain language model provides word embeddings that correspond to two-dimensional, real valued vectors. In this model, the words "vase" and "wall" have the vector representations $\vec{v}$ and $\vec{w}$, respectively

---

[2] An alternative introductory treatment of tensor calculus can be found in [8].

$$\vec{v} = (0, 1), \quad \text{and} \quad \vec{w} = (1, 0). \tag{2}$$

This representation could mean that they are context words in a count-based model, since they form the standard (orthogonal) basis of $\mathbb{R}^2$, or that they have this particular representation in a particular context-dependent language model. To compute cosine similarity, the notion of Euclidean inner product is used, where the components corresponding to a certain index are multiplied:

$$\vec{v} \cdot \vec{w} = 0 \cdot 1 + 1 \cdot 0 = 0, \tag{3}$$

which we can use to calculate the cosine of the angle $\theta$ between these vectors,

$$\cos(\theta) = \frac{\vec{v} \cdot \vec{w}}{\|\vec{v}\| \cdot \|\vec{w}\|} = \frac{0 \cdot 1 + 1 \cdot 0 = 0}{1 \cdot 1} = 0. \tag{4}$$

Thus, if the representations of these words are orthogonal, then using this measure to evaluate similarity we conclude that these words are not related. However, there is a degree of variation in the vectors that are assigned to the distributional semantics of each word. *Static embeddings* are unique vector representations given by a global analysis of a word over a corpus. The unique vector assigned to the semantics of a word depends on the model used to analyze the data, so different models do not necessarily put out the same vector representations. Alternative to this are *dynamic embeddings*, which assign different vector representations to the same word depending on context, within the same model.

Therefore, there are at least three ways in which the result of eq.4 and subsequent interpretation can be challenged:

1. **Static Embeddings.** If the representations come from a count-based model, choosing other words as context words changes the vector representation and therefore these words are not orthogonal to one another anymore; in fact this can happen with any static embedding representation when the basis of the representation changes. Examples of models that give static embeddings are Word2Vec [15] and GloVe [20].

2. **Dynamic Embeddings.** When the vector representations come from a context-dependent embedding, changing the context in which the words are evaluated influences their representation, which might not be orthogonal anymore. Dynamic embeddings can be obtained with i.e. ELMo [21] and BERT [7].

3. **Expectation of meaning.** Human judgements, which are the outcomes of experiments where subjects are explicitly asked to rate the similarity of words, predict that some words should have a degree of relationship. Therefore, the conclusion we derive from orthogonal representations of certain words might not be valid if there is a disagreement with their human assessment. These judgements are condensed in datasets such as the MEN dataset [3].

While points 1 and 2 can be related, caution is necessary in establishing that link. On a preliminary inspection, comparing the cosine similarity of context-free embeddings of nouns extracted from pre-trained BERT [7] with the normalized human judgements from the MEN dataset [3], we find that the similarity between two words given by the language model is systematically overrated when compared to its human counterpart. One possible explanation is that the language model is comparing all words against one another, so it is an important part of similarity that the two words belong to the the same part of speech, namely nouns, while humans assume that as a condition for similarity evaluation. Further, though we can ask the language model to rate the similarity of words in specific contexts, that has not explicitly been done with human subjects. A more detailed comparison between context-depend representations and human judgement constitutes further research.

One way to reconcile the variability of representations and the notion of similarity is to expand the notion of inner product to be invariant under the change of representations. Suppose now that by points 1 or 2 the representations of "vase" and "wall" change, respectively, to

$$\vec{v}' = (1,1),\ \vec{w}' = (-1,2). \tag{5}$$

These vectors also form a basis of $\mathbb{R}^2$, but not an orthogonal one. If we use the same measure to compute similarity, taking normalization into account, the Euclidean inner product gives $\vec{v}' \cdot \vec{w}' = (-1) \cdot 1 + 1 \cdot 2 = 1$ and cosine similarity gives

$$cos(\theta') = \frac{\vec{v}' \cdot \vec{w}'}{\|\vec{v}'\| \cdot \|\vec{w}'\|} = \frac{1}{\sqrt{2} \cdot \sqrt{5}} = \frac{1}{\sqrt{10}}. \tag{6}$$

If now we have a conflict between which representations are the correct ones, we can look at the human evaluations of similarity. Suppose that it corresponds too to $\frac{1}{\sqrt{10}}$.

We argue in this paper that, by introducing a different notion of inner product, we can fine-tune a relationship between the components of the vectors with the goal to preserve a particular value, for example a human similarity judgement. In this framework, the different representations of words in dynamic embeddings are brought about by a change of basis, similarly to what happens when the context words change in static embeddings, in which case the value of the inner product should be preserved. This can be achieved by describing the inner product as a tensor contraction between a vector and a dual vector, with the latter computed using a metric.

Let $V$ be a finite dimensional vector space and let $V^*$ denote its dual vector space, constituted by the linear maps from $V$ to the field $\mathbb{R}$. A tensor $T$ of type $(k, l)$ over $V$ is a multilinear map

$$T : \underbrace{V^* \times \cdots \times V^*}_{k} \times \underbrace{V \times \cdots \times V}_{l} \to \mathbb{R}. \qquad (7)$$

Once applied on $k$ dual vectors and $l$ vectors, a tensor outputs an element of the field, in this case a real number. By this token, a tensor of type $(0, 1)$ is a dual vector, which is the map from the vector space to the field, and a tensor of type $(1, 0)$, being technically the dual of a dual vector, is naturally isomorphic to a vector. Given a basis $E = \{\hat{e}_i\}$ in $V$ and its dual basis $_dE = \{\hat{e}^j\}$ in $V^*$, with $\hat{e}^j(\hat{e}_i) = \delta_i^j$, the tensor product between the basis vectors and dual basis vectors forms a basis $B = \{\hat{e}_{i_1} \otimes \cdots \otimes \hat{e}_{i_k} \otimes \hat{e}^{j_1} \otimes \cdots \otimes \hat{e}^{j_l}\}$ of a tensor of type $(k, l)$, allowing the tensor to be expressed with respect to this basis as

$$T = \sum_{i_1,\ldots,i_k,j_1,\ldots,j_l} T^{i_1\ldots i_k}{}_{j_1\ldots j_l} \hat{e}_{i_1} \otimes \cdots \otimes \hat{e}_{i_k} \otimes \hat{e}^{j_1} \otimes \cdots \otimes \hat{e}^{j_l}. \qquad (8)$$

The basis expansion coefficients $T^{i_1\ldots i_k}{}_{j_1\ldots j_l}$ are called the *components* of the tensor.

We can perform two important operations on tensors: apply the tensor product between them, $T' \otimes T$, and contract components of the tensor, $CT$. The first operation happens in the obvious way, while the second corresponds to applying one of the basis dual vectors to a basis vector, resulting in an identification and summing of the corresponding components:

$$(CT)^{i_1\ldots i_{k-1}}{}_{j_1\ldots j_{l-1}} = \sum_{\sigma} T^{i_1\ldots \sigma \ldots i_{k-1}}{}_{j_1\ldots \sigma \ldots j_{l-1}}. \qquad (9)$$

The outcome is a tensor of type $(k - 1, l - 1)$. Note that this procedure is basis independent, because of the relationship between the basis and dual basis. For a tensor of type $(1, 1)$, which represents a linear operator from $V$ to $V$, tensor contraction corresponds precisely to taking the trace of that operator. To simplify the notation, we will use primed indices instead of numbered ones when the tensors have a low rank. We define a special $(0, 2)$ tensor called a *metric* $d$:

$$d = \sum_{j,j'} d_{jj'} \hat{e}^j \otimes \hat{e}^{j'}. \tag{10}$$

This tensor is symmetric and non-degenerate. The contraction of this tensor with two vectors $v$ and $w$ gives the value of the inner product:

$$d(v,w) = \sum_{j,j'} d_{jj'} v^j w^{j'}. \tag{11}$$

Because of symmetry, $d(v,w) = d(w,v)$, and because of non-degeneracy, the metric is invertible, with its inverse $d^{-1}$ expressed as

$$d^{-1} = \sum_{i,i'} d^{ii'} \hat{e}_i \otimes \hat{e}_{i'}. \tag{12}$$

Given that the elements extracted from the data are elements of $V$, the contractions that need to be performed, for example for the application of the compositionality principle in vector spaces, must involve a passage from vectors to dual vectors as seen in the DisCoCat model, before contraction takes place. The metric can be used to define a canonical map between $V$ and $V^*$ via the partial map that is obtained when only one vector is used as an argument of the metric, giving rise to the dual vector $_d v : v \mapsto d(-,v)$, with the slash indicating the empty argument slot:

$$d(v,w) \equiv d(v,-)(w) \equiv {_d v}(w). \tag{13}$$

This formulation is basis independent, since it results from tensor contraction. Once a basis is defined, the resulting dual vector can be expressed as

$$v^d = \sum_{i,j,j'} d_{jj'} v^i \hat{e}^j \otimes \hat{e}^{j'}(\hat{e}_i) = \sum_{j,j'} d_{jj'} v^j \hat{e}^{j'} = \sum_{j'} v_{j'} \hat{e}^{j'}, \tag{14}$$

where we rewrite $v_{j'} = \sum_j d_{jj'} v^j$.

We call the components of vectors, with indices "up", the *contravariant* components, and those of dual vectors, with indices "down", the *covariant* components. Thus, consistent with our notation, the metric can be used to "lower" or "raise" indices, applying contraction between the metric and the tensor and relabeling the components:

$$d(T) = \sum_{i_1,\ldots,i_k,j_1,\ldots,j_{l+2}} d_{j_{l+1},j_{l+2}} T^{i_1,\ldots,i_k}{}_{j_1,\ldots,j_l} \hat{e}^{j_{l+1}} \otimes \hat{e}^{j_{l+2}}(\hat{e}_{i_1}) \otimes \ldots \otimes \hat{e}^{j_l}$$
$$= \sum_{i_1,\ldots,i_k,j_1,\ldots,j_{l+1}} d_{j_{l+1},i_1} T^{i_1,\ldots,i_k}{}_{j_1,\ldots,j_l} \hat{e}^{j_{l+1}} \otimes \hat{e}_{i_2} \otimes \ldots \otimes \hat{e}^{j_l}$$
$$= \sum_{i_2,\ldots,i_k,j_1,\ldots,j_{l+1}} T_{j_{l+1}}{}^{i_2,\ldots,i_k}{}_{j_1,\ldots,j_l} \hat{e}^{j_{l+1}} \otimes \hat{e}_{i_2} \otimes \ldots \otimes \hat{e}^{j_l}. \quad (15)$$

The effect of the metric on a tensor can be captured by seeing how we rewrite the components of some example tensors:

- $\sum_{j'} d_{jj'} T^{j'}{}_{j''} = T_{jj''}$;
- $\sum_{i'} T^i{}_{i'} d^{i'i''} = T^{ii''}$;
- $\sum_{j',j'''} d_{jj'} d_{j''j'''} T^{j'j'''} = T_{jj''}$.

Most importantly, a proper tensor is only defined in the form of eq.8, so whenever we have a tensor that has components "up" and "down" in different orders, for example in $T_j{}^i$, this is in fact a tensor of type $(1,1)$ of which the actual value of the components is

$$\sum_{i',j'} d^{ii'} d_{jj'} T^{j'}{}_{i'}. \quad (16)$$

Returning to our toy example with the words "vase" and "wall", we can look at the change in vector representations as a change of basis $\hat{e}_i = \sum_{i'} \Lambda_i{}^{i'} \hat{e}'_{i'}$:

$$\vec{v} = \sum_i v^i \hat{e}_i = \sum_{ii'} v^i \Lambda_i{}^{i'} \hat{e}'_{i'} = \sum_{i'} v'^{i'} \hat{e}'_{i'}, \quad (17)$$

corresponding to a change in the vector components $v'^{i'} = v^i \Lambda_i{}^{i'}$. The components of the metric also change with the basis:

$$d'_{j''j'''} = \Lambda_{j'''}{}^{j'} \Lambda_{j''}{}^{j} d_{jj'}. \quad (18)$$

With this change, we can show that inner product remains invariant under a basis change:

$$w'^{i'} v'_{i'} = w'^{i'} v'^{j'} d'_{j'i'} = w'^{i'} v'^{j'} \Lambda_i{}^i \Lambda_j{}^j d_{ji} = w^i v^j d_{ji} = w^i v_i. \quad (19)$$

In this way, finding the right metric allows us to preserve a value that is constant in the face of context dependent representations. Assuming a metric that has the following matrix representation in the standard basis,

$$d = \begin{pmatrix} 2 & 1 \\ 1 & 5 \end{pmatrix}, \tag{20}$$

its application to the vector elements in eqs.2 gives a value of the inner product calculated in the new representation:

$$v'_{i'} w'^{i'} = \begin{pmatrix} 1 & 0 \end{pmatrix} \begin{pmatrix} 2 & 1 \\ 1 & 5 \end{pmatrix} \begin{pmatrix} 0 \\ 1 \end{pmatrix} = 1. \tag{21}$$

Since norms of the vectors have to be calculated using the same notion of inner product,

$$\|\vec{v}\| = \sqrt{v^i g_{ij} v^j}, \tag{22}$$

we find exactly the cosine similarity calculated in eq.6. Note that this formalism allows us to deal with non-orthogonal basis, but does not require it: in fact, there is an implicit metric already when we compute the Euclidean inner product in eq.2, given by $d_{orth} = \begin{pmatrix} 1 & 0 \\ 0 & 1 \end{pmatrix}$ in the standard basis, which is the one assumed when talking about an orthonormal basis.

Since these new tools allow us to preserve a quantity in the face of a change of representation, we can start reversing the question on similarity: given a certain human judgement on similarity, or another constant of interest, what is the metric that preserves it across different representations?[3] Once the vector spaces are endowed with specific metrics, the new inner product definitions permeate all higher-rank tensor contractions that are performed between higher and lower rank tensors, namely the ones that will be used in the interpretation of the Lambek rules,[4]

---

[3] In case the quantity we wish to preserve is other than that of the Euclidean inner product in either representation, there is an option to expand the vector representation of our words by adding vector components that act as parameters, to ensure that the quantity is indeed conserved. This would be similar to the role played by the time dimension in Einstein's relativity theory.

[4] Using this formalism, we can replace the unit and counit maps $\epsilon$ and $\eta$ maps of the compact closed category **FdVect** by

$$\eta^l : \mathbb{R} \to V \otimes V^* :: 1 \mapsto \mathbb{1} \otimes d(\mathbb{1}, -)$$
$$\eta^r : \mathbb{R} \to V^* \otimes V :: 1 \mapsto d(-, \mathbb{1}) \otimes \mathbb{1}$$
$$\epsilon^l : V^* \otimes V \to \mathbb{R} :: d(-, v) \otimes u \mapsto d(u, v)$$
$$\epsilon^r : V \otimes V^* \to \mathbb{R} :: v \otimes d(u, -) \mapsto d(u, v).$$

and can further be extended to density matrices.

## 3.1 Metric in Dirac Notation

We want to lift our description to the realm of density matrices. We now show how the concept of a metric can also be introduced in that description, such that the previously described advantages carry over.

Dirac notation is the usual notation for vectors in the quantum mechanics literature. To make the bridge with the previous concepts from tensor calculus, we introduce it simply as a different way to represent the basis and dual basis of a vector space. Let us rename their elements as *kets* $|i\rangle \equiv \hat{e}_i$ and as *bras* $\langle j| \equiv \hat{e}^j$. The fact that the bases are dual to one another is expressed by the orthogonality condition $\langle j|i\rangle = \delta_{ij}$, which, if the vector basis elements are orthogonal to each other, is equivalent to applying the Euclidean metric to $|i\rangle$ and $|j\rangle$. Using Dirac notation, a vector and dual vector are represented as $v \equiv |v\rangle = \sum_i v^i |i\rangle$ and $v^d \equiv \langle u| = \sum_j v_j \langle j|$.[5] If the basis elements are not orthogonal, this mapping has to be done through a more involved metric. To express this, in this paper we introduce a modified Dirac notation over the field of real numbers, inspired by the one used in [9] for the treatment of quantum states related by a specific group structure.[6] The previous basis elements of $V$ are written now as $|_i\rangle \equiv \hat{e}_i$ and the corresponding dual basis as $\langle^j| \equiv \hat{e}^j$, such that $\langle^j|_i\rangle = \delta_i^j$. In this basis, the metric is expanded as $d = \sum_{j,j'} d_{j'j} \langle^j| \otimes \langle^{j'}|$ while the inverse metric is expressed as $d^{-1} = \sum_{ii'} d^{i'i} |_i\rangle \otimes |_{i'}\rangle$. The elements of the metric and inverse metric are related by $\sum_i d_{j'i'} d^{i'i} = \delta_{j'}^{i'}$. Applying the metric to a basis element of $V$, we get

$$\langle_i| \equiv d(-, |_i\rangle) = \sum_{jj'} d_{j'j} \langle^j| \otimes \langle^{j'}|_i\rangle = \sum_j d_{ij} \langle^j|. \qquad (23)$$

Acting with this on $|_{i'}\rangle$ to extract the value of the inner product, the following formulations are equivalent:

$$d(|_{i'}\rangle, |_i\rangle) = d(-, |_i\rangle) |_{i'}\rangle = \sum_j d_{ij} \langle^j|_{i'}\rangle = \langle_i|_{i'}\rangle = d_{ii'}. \qquad (24)$$

When the inverse metric is applied to $\langle^j|$ it gives

---

[5] For orthonormal basis over the field of complex numbers, the covariant components are simply given by the complex conjugate of the contravariant ones, $v_i = \bar{v}^i$.

[6] This treatment can be extended to the field of complex numbers by considering that the metric has conjugate symmetry, $d_{ij} = \bar{d}_{ji}$ [25].

$$\left|^{j}\right\rangle \equiv d\left(-,\left\langle ^{j}\right|\right) = \left\langle ^{j}\right| \sum_{ii'} d^{i'i} \left|_{i}\right\rangle \otimes \left|_{i'}\right\rangle = \sum_{i'} d^{i'j} \left|_{i'}\right\rangle, \qquad (25)$$

with a subsequent application on $\left\langle ^{j'}\right|$ giving

$$d^{-1}\left(\left\langle ^{j'}\right|,\left\langle ^{j}\right|\right) = \left\langle ^{j'}\right| d\left(-,\left\langle ^{j}\right|\right) = \left\langle ^{j'}\right| \sum_{i'} d^{i'j} \left|_{i'}\right\rangle = \left\langle ^{j'}\right|^{j}\right\rangle = d^{j'j}. \qquad (26)$$

Consistently, we can calculate the value of the new bras and kets defined in eqs.23 and 25 applied to one other, showing that they too form a basis/dual basis pair:

$$\left\langle _{i}\right|^{j}\right\rangle = \sum_{j'} d_{ij'} \left\langle ^{j'}\right| \sum_{i'} d^{i'j} \left|_{i'}\right\rangle = \sum_{i'j'} d_{ij'} d^{i'j} \left\langle ^{j'}\right|_{i'}\right\rangle = \sum_{j'} d_{ij'} d^{j'j} = \delta_{i}^{j}. \qquad (27)$$

If the basis elements are orthogonal, the components of the metric and inverse metric coincide with the orthogonality condition.

## 4 Density Matrices: Capturing Directionality

The semantic spaces we envisage for the interpretation of the syntactic calculus are density matrices. A *density matrix* or density operator is used in quantum mechanics to describe systems for which the state is not completely known. For lexical semantics, it can be used to describe the meaning of a word by placing distributional information on its components. As standardly presented,[7] density matrices that are defined on a tensor product space indicate no preference with respect to contraction from the left or from the right. Because we want to keep the distinction between left and right implications in the semantics, we set up the interpretation of composite spaces in such a way that they indicate which parts will and will not contract with other density matrices.

The *basic* building blocks of the interpretation are density matrix spaces $\tilde{V} \equiv V \otimes V^{*}$. For this composite space, we choose the basis formed by $\left|_{i}\right\rangle$ tensored with $\left\langle _{i'}\right|$, $\tilde{E} = \{\left|_{i}\right\rangle \left\langle _{i'}\right|\} = \{\tilde{E}_{J}\}$. Carrying over the notion of duality to the density matrix space, we define the dual density matrix space $\tilde{V}^{*} \equiv V \otimes V^{*}$. The dual basis in this space is the map that takes each basis element of $\tilde{V}$ and returns the appropriate orthogonality conditions. It is formed by $\left\langle ^{j}\right|$ tensored with $\left|^{j'}\right\rangle$, $_{d}\tilde{E} = \{\left|^{j'}\right\rangle \left\langle ^{j}\right|\} = \{\tilde{E}^{J}\}$, and is applied on the basis vectors of $\tilde{V}$ via the trace operation

---

[7] A background for the non-physics reader can be found in [19].

$$\tilde{E}^J\left(\tilde{E}_I\right) = \text{Tr}\left(|i\rangle\langle_{i'}|^{j'}\rangle\langle^j|\right) = \sum_l \langle^l|i\rangle\langle_{i'}|^{j'}\rangle\langle^j|l\rangle$$
$$= \sum_{jj'}\langle^j|i\rangle\langle_{j'}|^{i'}\rangle \delta^j_i \delta^{j'}_{i'} \equiv \delta^J_I. \tag{28}$$

Because density operators are hermitian, their matrices do not change under conjugate transposition, which extends to elements of the basis of the density matrix space. In this way, we can extend our notion of metric to the space of density matrices, where a new metric $D$ emerges from $d$, expanded in the basis of $V^*$ as

$$D = \sum_{J,J'} D_{JJ'} \tilde{E}^J \otimes \tilde{E}^{J'} \tag{29}$$

$$= \sum_{jj',j''j'''} d_{j''j'} d_{j'''j} \left|^{j'}\right\rangle\left\langle^j\right| \otimes \left|^{j'''}\right\rangle\left\langle^{j''}\right|. \tag{30}$$

We can see how both definitions are equivalent by their action on a density matrix tensor $T \equiv \sum_I T^I \tilde{E}_I \equiv \sum_{ii'} T^{ii'} |i\rangle\langle_{i'}|$. Staying at the level of $\tilde{V}$ and $\tilde{V}^*$, we use eq.29 to obtain

$$D(-,T) = \sum_{I,J,J'} D_{JJ'} T^I \tilde{E}^J \otimes \tilde{E}^{J'}\left(\tilde{E}_I\right) = \sum_{I,J,J'} D_{JJ'} T^I \tilde{E}^J \delta^{J'}_I$$
$$= \sum_{J,J'} D_{JJ'} T^{J'} \tilde{E}^J \equiv \sum_J T_J \tilde{E}^J = \sum_{jj'} T_{j'j} \left|^{j'}\right\rangle\left\langle^j\right|, \tag{31}$$

where we redefine $T_J \equiv D_{JJ'} T^{J'}$, thus establishing covariance and contravariance of the tensor components defined over the density matrix space. Looking in its turn at the level of $V$ and $V^*$, using eq.30, we see that both definitions are equivalent:

$$D(-,T) = \sum_{ii',jj',j''j'''} T^{ii'} d_{j''j'} d_{j'''j} \left|^{j'}\right\rangle\left\langle^j\right| \otimes \text{Tr}\left(\left|^{j'''}\right\rangle\left\langle^{j''}\right|i\rangle\langle_{i'}|\right)$$
$$= \sum_{ii',jj',j''j'''} T^{ii'} d_{j''j'} d_{j'''j} \delta^{j''}_i \delta^{j'''}_{i'} \left|^{j'}\right\rangle\left\langle^j\right|$$
$$= \sum_{ii'jj'} T^{ii'} d_{ij'} d_{i'j} \left|^{j'}\right\rangle\left\langle^j\right| \equiv \sum_{jj'} T_{jj'} \left|^{j'}\right\rangle\left\langle^j\right|, \tag{32}$$

where we rewrite $T_{jj'} \equiv T^{ii'} d_{ij'} d_{i'j}$.[8]

From these basic building blocks, *composite* spaces are formed via the binary operation $\otimes$ (tensor product) and a unary operation $()^*$ (dual functor) that sends the elements of a density matrix basis to its dual basis, using the metric defined above. In the notation, we use $\tilde{A}$ for density matrix spaces (basic or compound), and $\rho$, or subscripted $\rho_x, \rho_y, \rho_z, \ldots \in \tilde{A}$ for elements of such spaces. The $()^*$ operation is involutive; it interacts with the tensor product as $(\tilde{A} \otimes \tilde{B})^* = \tilde{B}^* \otimes \tilde{A}^*$ and acts as identity on matrix multiplication.

Below in (†) is the general form of a density matrix defined on a single space in the standard basis, and (‡) in the dual basis:

$$(\dagger) \quad \rho_x^{\tilde{A}} = \sum_{ii'} X^{ii'} \ket{i}_{\tilde{A}} \bra{i'}, \qquad (\ddagger) \quad \rho_x^{\tilde{A}^*} = \sum_{jj'} X_{j'j} \ket{j'}_{\tilde{A}^*} \bra{j}.$$

Over the density matrix spaces, we can see these matrices as *tensors* as we defined them previously, with $X^I \equiv X^{ii'}$ the *contravariant* components and with $X_{J'} \equiv X_{j'j}$ the *covariant* components.

A density matrix of a composite space can be an element of the tensor product space between the standard space and the dual space either from the left or from the right:

$$\rho_y^{\tilde{A} \otimes \tilde{B}^*} = \sum_{ii',jj'} Y^{ii'}_{j'j} \ket{i \, j'}_{\tilde{A} \otimes \tilde{B}^*} \bra{i' \, j}; \tag{33}$$

$$\rho_w^{\tilde{B}^* \otimes \tilde{A}} = \sum_{ii',jj'} W_{j'j}^{\;\;ii'} \ket{j' \, i}_{\tilde{B}^* \otimes \tilde{A}} \bra{j \, i'}. \tag{34}$$

Although both tensors are of the form $(1,1)$, the last one is a tensor with components $Y^I_{J'}$, which relate with a true tensor form by $D^{II'} Y_{I'}^{\;\;J} D_{JJ'}$. Recursively, density matrices that live in higher-rank tensor product spaces can be constructed, taking a tensor product with the dual basis either from the left or from the right. Multiplication between two density matrices of a standard and a dual space follows the rules of tensor contraction:

$$\rho_y^{\tilde{A}^*} \cdot \rho_x^{\tilde{A}} = \sum_{jj'} Y_{j'j} \ket{j'}_{\tilde{A}^*} \bra{j} \cdot \sum_{ii'} X^{ii'} \ket{i}_{\tilde{A}} \bra{i'} = \sum_{i',jj'} Y_{j'j} X^{ji'} \ket{j'}_{\tilde{A}} \bra{i'}. \tag{35}$$

---

[8] Here we can compare our formalism to that of the compact closed category of completely positive maps **CPM(FdVect)** developed in [26]. The categorical treatment applies here at a higher level, however, introducing the metric defines explicitely the canonical isomorphisms $V \cong V^*$ and $\tilde{V} \cong \tilde{V}^*$, which trickles down to knowing exactly how the symmetry of the tensor product acts on the compenents of a tensor: $\sigma_{V,V^*} : V^* \otimes V \to V \otimes V^* :: \sum_{ij} T_i^{\;\;j} \hat{e}^i \otimes \hat{e}_j \mapsto \sum_{ii',jj'} d^{ii'} d_{jj'} T^{j'}_{\;\;i'} \hat{e}_i \otimes \hat{e}^j$.

$$\rho_x^{\tilde{A}} \cdot \rho_y^{\tilde{A}^*} = \sum_{ii'} X^{ii'} |i\rangle_{\tilde{A}} \langle i'| \cdot \sum_{jj'} Y_{j'j} |j'\rangle_{\tilde{A}^*} \langle j| = \sum_{i,jj'} X^{ij'} Y_{j'j} |i\rangle_{\tilde{A}} \langle j|, \quad (36)$$

respecting the directionality of composition. To achieve full contraction, the trace in the appropriate space is applied, corresponding to a partial trace if the tensors involve more spaces:

$$\operatorname{Tr}_{\tilde{A}} \left( \sum_{i',jj'} Y_{j'j} X^{ji'} |j'\rangle_{\tilde{A}} \langle i'| \right) = \sum_{l,i',jj'} Y_{j'j} X^{ji'} {}_{\tilde{A}}\langle l|j'\rangle_{\tilde{A}^*} {}_{\tilde{A}}\langle i'|l\rangle_{\tilde{A}^*} = \sum_{jj'} Y_{j'j} X^{jj'}, \quad (37)$$

$$\operatorname{Tr}_{\tilde{A}} \left( \sum_{i,jj'} X^{ij'} Y_{j'j} |i\rangle_{\tilde{A}} \langle j| \right) = \sum_{l,j',ij} X^{ij'} Y_{j'j} {}_{\tilde{A}^*}\langle l|i\rangle_{\tilde{A}} {}_{\tilde{A}^*}\langle j|l\rangle_{\tilde{A}} = \sum_{jj'} X^{jj'} Y_{j'j}. \quad (38)$$

We see that the cyclic property of the trace is preserved.

In §6 we will be dealing with derivational ambiguity, and for that the concepts of *subsystem* and *permutation operation* introduced here will be useful. A subsystem can be thought of as a copy of a space, described using the same basis, but formally treated as a different space. In practice, this means that different subsystems do not interact with one another. In the quantum setting, they represent independent identical quantum systems. For example, when we want to describe the spin states of two electrons, despite the fact that each spin state is defined on the same basis, it is necessary to distinguish which electron is in which state and so each is attributed to their own subsystem. Starting from a space $\tilde{A}$, two different subsystems are referred to as $\tilde{A}_1$ and $\tilde{A}_2$. If different words are described in the same space, subsystems can be used to formally assign them to different spaces. The permutation operation extends naturally from the one in standard quantum mechanics. We define two permutation operators: $P^{\tilde{A}_1 \tilde{A}_2}$ permutes the elements of the basis of the respective spaces, while $P_{\tilde{A}_1 \tilde{A}_2}$ permutes the elements of the dual basis. If only one set of basis elements is inside the scope of the permutation operators, then either the subsystem assignment changes,

$$P^{\tilde{A}_1 \tilde{A}_2} |i\rangle_{\tilde{A}_1} \langle i'| P^{\tilde{A}_1 \tilde{A}_2} = |i\rangle_{\tilde{A}_2} \langle i'|; \qquad P_{\tilde{A}_1 \tilde{A}_2} |i'\rangle_{\tilde{A}_1^*} \langle i| P_{\tilde{A}_1 \tilde{A}_2} = |i'\rangle_{\tilde{A}_2^*} \langle i|; \quad (39)$$

or the respective space of tracing changes,

$$\operatorname{Tr}_{\tilde{A}_1}\left(P^{\tilde{A}_1\tilde{A}_2}|i'\rangle_{\tilde{A}_2^*}\langle i|P^{\tilde{A}_1\tilde{A}_2}\right) = \operatorname{Tr}_{\tilde{A}_2}\left(|i'\rangle_{\tilde{A}_2^*}\langle i|\right). \tag{40}$$

Note that permutations take precedence over traces. If two words are assigned to different subsystems, the permutations act to swap their space assignment:[9]

$$P^{\tilde{A}_1\tilde{A}_2}|i\rangle_{\tilde{A}_1}\langle i'| \otimes |j\rangle_{\tilde{A}_2}\langle j'|P^{\tilde{A}_1\tilde{A}_2} = |i\rangle_{\tilde{A}_2}\langle i'| \otimes |j\rangle_{\tilde{A}_1}\langle j'|, \tag{41}$$

$$P_{\tilde{A}_1\tilde{A}_2}|i'\rangle_{\tilde{A}_1^*}\langle i| \otimes |j'\rangle_{\tilde{A}_2^*}\langle j|P_{\tilde{A}_1\tilde{A}_2} = |i'\rangle_{\tilde{A}_2^*}\langle i| \otimes |j'\rangle_{\tilde{A}_1^*}\langle j|. \tag{42}$$

If no word has that subsystem assignment then the permutation has no effect.

## 5 Interpreting Lambek Calculus derivations

Let us turn now to the syntax-semantics interface, which takes the form of a homomorphism sending the types and derivations of the syntactic front end $(\mathbf{N})\mathbf{L}_{/,\backslash}$ to their semantic counterparts. Consider first the action of the interpretation homomorphism on *types*. We write $\lceil \cdot \rceil$ for the map that sends syntactic types to the interpreting semantic spaces. For primitive types we set

$$\lceil s \rceil = \tilde{S}, \ \lceil np \rceil = \lceil n \rceil = \tilde{N}, \tag{43}$$

with $S$ the vector space for sentence meanings and $N$ the space for nominal expressions (common nouns, full noun phrases). For compound types we have

$$\lceil A/B \rceil = \lceil A \rceil \otimes \lceil B \rceil^*, \text{ and } \lceil A\backslash B \rceil = \lceil A \rceil^* \otimes \lceil B \rceil. \tag{44}$$

Given semantic spaces for the syntactic types, we can turn to the interpretation of the syntactic *derivations*, as coded by their $\lambda_{/,\backslash}$ proof terms. We write $[\![\cdot]\!]_g$ for the map that associates each term $t$ of type $A$ with a semantic value, i.e. an element of $\lceil A \rceil$, the semantic space where meanings of type $A$ live. The map $[\![\cdot]\!]$ is defined relative to a assignment function $g$ that provides a semantic value for the basic building blocks, viz. the variables that label the axiom leaves of a proof. As we saw

---

[9]We define this as a shorthand application of the permutation operations as defined in eq.39, such that eq.41 can be calculated w.r.t. that definition as

$$P^{\tilde{A}_1\tilde{A}_2}|i\rangle_{\tilde{A}_1}\left(_{\tilde{A}_1}\langle i'|P^{\tilde{A}_1\tilde{A}_2}\right) \otimes \left(P^{\tilde{A}_1\tilde{A}_2}|j\rangle_{\tilde{A}_2}\right)_{\tilde{A}_2}\langle j'|P^{\tilde{A}_1\tilde{A}_2}$$

$$= P^{\tilde{A}_1\tilde{A}_2}|i\rangle_{\tilde{A}_1}{}_{\tilde{A}_2}\langle i'| \otimes |j\rangle_{\tilde{A}_1}{}_{\tilde{A}_2}\langle j'|P^{\tilde{A}_1\tilde{A}_2} = |i\rangle_{\tilde{A}_2}\langle i'| \otimes |j\rangle_{\tilde{A}_1}\langle j'|,$$

and similarly for eq.42.

above, a proof term is a generic meaning recipe that abstracts from particular lexical meanings. Specific lexical items, as we will see in §6, have the status of *constants*. These constants are mapped to their distributional meaning by an interpretation function $I$. The distributional meaning corresponds to the embeddings assigned by a particular model to the lexicon. Below we show that this calculus is sound with respect to the semantics of section 4.

**Axiom**

$$[\![x^A]\!]_g = g(x^A) = \rho_x^{\lceil A \rceil} = \sum_{ii'} X^{ii'} |i\rangle_{\lceil A \rceil} \langle i'|. \tag{45}$$

**Elimination**  Recall the inference rules of fig.1
$E_/$: Premises $t^{B/A}$, $u^A$; conclusion $(t \triangleleft u)^B$:

$$[\![(t \triangleleft u)^B]\!]_g \equiv \mathrm{Tr}_{\lceil A \rceil} \left( [\![t^{B/A}]\!]_g \cdot [\![u^A]\!]_g \right) \tag{46}$$

$$= \mathrm{Tr}_{\lceil A \rceil} \left( \sum_{ii',jj'} T^{ii'}_{j'j} \left| {j' \atop i} \right\rangle_{\lceil B \rceil \otimes \lceil A \rceil^*} \left\langle {j \atop i'} \right| \cdot \sum_{kk'} U^{kk'} |k\rangle_{\lceil A \rceil} \langle k'| \right) \tag{47}$$

$$= \sum_{ii',jj'} \sum_{kk'} T^{ii'}_{j'j} \cdot U^{kk'} \delta^j_k \delta^{j'}_{k'} |i\rangle_{\lceil B \rceil} \langle i'| = \sum_{ii',jj'} T^{ii'}_{j'j} \cdot U^{jj'} |i\rangle_{\lceil B \rceil} \langle i'|. \tag{48}$$

$E_\backslash$: Premises $u^A$, $t^{A\backslash B}$; conclusion $(u \triangleright t)^B$:

$$[\![(u \triangleright t)^B]\!]_g \equiv \mathrm{Tr}_{\lceil A \rceil} \left( [\![u^A]\!]_g \cdot [\![t^{A\backslash B}]\!]_g \right) \tag{49}$$

$$= \mathrm{Tr}_{\lceil A \rceil} \left( \sum_{kk'} U^{kk'} |k\rangle_{\lceil A \rceil} \langle k'| \cdot \sum_{ii',jj'} T_{jj'}^{ii'} \left| {j' \atop i} \right\rangle_{\lceil A \rceil^* \otimes \lceil B \rceil} \left\langle {j \atop i'} \right| \right) = \tag{50}$$

$$= \sum_{kk'} \sum_{ii',jj'} U^{kk'} \cdot T^{ii'}_{j'j} \delta^j_k \delta^{j'}_{k'} |i\rangle_{\lceil B \rceil} \langle i'| = \sum_{ii',jj'} U^{jj'} \cdot T_{j'j}^{ii'} |i\rangle_{\lceil B \rceil} \langle i'|. \tag{51}$$

**Introduction**  $I_/$: Premise $t^B$, with $x^A$ as its rightmost parameter; conclusion $(\lambda^r x.t)^{B/A}$:

$$[\![(\lambda^r x.t)^{B/A}]\!]_g \equiv \sum_{kk'} \left( [\![t^B]\!]_{g^x_{kk'}} \otimes |k'\rangle_{\lceil A \rceil^*} \langle k| \right) \tag{52}$$

$I_{\backslash}$: Premise $t^B$, with $x^A$ as its leftmost parameter; conclusion $(\lambda^l x.t)^{A\backslash B}$:

$$\left[\!\left[\left(\lambda^l x.t\right)^{A\backslash B}\right]\!\right]_g \equiv \sum_{kk'} \left( \left|k'\right\rangle_{\lceil A \rceil^*} \left\langle k\right| \otimes [\![t^B]\!]_{g^x_{kk'}} \right) \tag{53}$$

Here $g^x_{kk'}$ is the assignment exactly like $g$ except possibly for the parametric variable $x$ which takes the value of the basis element $|k\rangle_{\lceil A \rceil}\langle k'|$. More generally, the interpretation of the introduction rules lives in a compound density matrix space representing a linear map from $\tilde{A}$ to $\tilde{B}$. The semantic value of that map, applied to any object $m \in \tilde{A}$, is given by $[\![t^B]\!]_{g'}$, where $g'$ is the assignment exactly like $g$ except possibly for the bound variable $x^A$, which is assigned the value $m$. Note that now, given the introduction of the metric, the interpretations of $A/B$ and $B\backslash A$ are related by it: if the components of the first are $T_J{}^I$, then those of the second are given by those in eq.16 adapted for density matrices. This is what introduces directionality in our interpretation: using the metric, we can extract a certain representation for a function word and distinguish by the values of the components whether it will contract from the left or from the right.

## 6 Derivational Ambiguity

The density matrix construction has been successfully used to address lexical ambiguity [22], as well as lexical and sentence entailment [1,24], where different measures of entropy are used to perform the disambiguation. Here we arrive at disambiguation in a different way, by storing in the diagonal elements of a higher order density matrix the different interpretations that result from the different contractions that the proof-as-programs prescribes. This is possible due to the the set-up that is formed by a multi-partite density matrices space, so that, by making use of permutation operations, it happens automatically that the two meanings are expressed independently. This is useful because it can be integrated with a lexical interpretation in density matrices optimized to other tasks, such as lexical ambiguity or entailment. It is also appropriate to treat the existence of these ambiguities in the context of incrementality, since it keeps the meanings separated in their interaction with posterior fragments.

We give a simple example of how the trace machinery can be used on an ambiguous fragment, providing a passage from one reading to the other at the interpretation level, and how the descriptions are kept separated. For this application, the coefficients in the interpretation of the words contain distributional information harvested from data, either from a count-base model or a more sophisticated language model.

The final coefficient of each outcomes is the vector-based representation of that reading.

We illustrate the construction with the phrase "tall person from Spain". The lexicon below has the syntactic type assignments and the corresponding semantic spaces.

|        | syn type $A$   | $\lceil A \rceil$                                      |
|-------:|:--------------:|:-------------------------------------------------------|
| tall   | $n/n$          | $N^* \otimes N \otimes (N^* \otimes N)^*$              |
| person | $n$            | $N^* \otimes N$                                        |
| from   | $(n\backslash n)/np$ | $(N^* \otimes N)^* \otimes N^* \otimes N \otimes (N^* \otimes N)^*$ |
| Spain  | $np$           | $N^* \otimes N$                                        |

Given this lexicon, "tall person from Spain" has two derivations, corresponding to the bracketings "(tall person) from Spain" ($x$/tall, $y$/person, $w$/from, $z$/Spain):

$$\cfrac{\cfrac{\overline{x:n/n \vdash x:n/n}^{ax} \quad \overline{y:n \vdash y:n}^{ax}}{(x:n/n, y:n) \vdash (x \triangleleft y) : n} /_{E_2} \quad \cfrac{\overline{w:(n\backslash n)/np \vdash w:(n\backslash n)/np}^{ax} \quad \overline{z:np \vdash z:np}^{ax}}{(w:(n\backslash n)/np, z:n) \vdash (w \triangleleft z) : n\backslash n} /_{E_1}}{[(x:n/n, y:n), (w:(n\backslash n)/np, z:n)] \vdash ((x \triangleleft y) \triangleright (w \triangleleft z)) : n} \backslash_{E_3}$$

versus "tall (person from Spain)":

$$\cfrac{\overline{x:n/n \vdash x:n/n}^{ax} \quad \cfrac{\overline{y:n \vdash y:n}^{ax} \quad \cfrac{\overline{w:(n\backslash n)/np \vdash w:(n\backslash n)/np}^{ax} \quad \overline{z:np \vdash z:np}^{ax}}{(w:(n\backslash n)/np, z:n) \vdash (w \triangleleft z) : n\backslash n} /_{E_1}}{[y:n, (w:(n\backslash n)/np, z:n)] \vdash (y \triangleright (w \triangleleft z)) : n} \backslash_{E_2}}{(x:n/n, [y:n, (w:(n\backslash n)/np, z:n)]) \vdash (x \triangleleft (y \triangleright (w \triangleleft z))) : n} /_{E_3}$$

In the first reading, the adjective "tall" is evaluated with respect to all people, before it is specified that this person happens to be from Spain, whereas in the second reading the adjective "tall" is evaluated only in the restricted universe of people from Spain.

Taking "from Spain" as a unit for simplicity, let us start with the following primitive interpretations:

- $[\![tall^{n/n}]\!]_I = \sum_{ii',jj'} \mathbf{T}_{ii'}^{j'j} \left|{i \atop j'}\right\rangle_{N \otimes N^*} \left\langle{i' \atop j}\right|,$

- $[\![person^n]\!]_I = \sum_{kk'} \mathbf{P}_{kk'} \left|k\right\rangle_N \left\langle k'\right|,$

- $[\![from\_Spain^{n\backslash n}]\!]_I = \sum_{ll',mm'} \mathbf{F}^{l'l}_{mm'} \left|{m \atop l'}\right\rangle_{N^* \otimes N} \left\langle{m' \atop l}\right|.$

Interpreting each step of the derivation in the way described in the previous section will give two different outcomes. The first one is

$$[\![tall\_person\_from\_Spain^n]\!]_I^1 =$$
$$= \text{Tr}_N \left( \text{Tr}_N \left( \sum_{ii',jj'} \mathbf{T}^{ii'}_{j'j} \left|{}^{j'}_{i}\right\rangle_{N\otimes N^*} \left\langle{}^{j}_{i'}\right| \cdot \sum_{kk'} \mathbf{P}^{kk'} |k\rangle_N \langle k'| \right) \right.$$
$$\left. \cdot \sum_{ll',mm'} \mathbf{F}^{mm'}_{l'l} \left|{}^{l'}_{m}\right\rangle_{N^*\otimes N} \left\langle{}^{l}_{m'}\right| \right)$$
$$= \sum_{ii',jj',mm'} \mathbf{T}^{ii'}_{j'j} \mathbf{P}^{jj'} \mathbf{F}^{mm'}_{i'i} |m\rangle_N \langle m'|, \qquad (54)$$

while the second one is

$$[\![tall\_person\_from\_Spain^n]\!]_I^2 =$$
$$= \text{Tr}_N \left( \sum_{ii',jj'} \mathbf{T}^{ii'}_{j'j} \left|{}^{j'}_{i}\right\rangle_{N\otimes N^*} \left\langle{}^{j}_{i'}\right| \cdot \text{Tr}_N \left( \sum_{kk'} \mathbf{P}_{kk'} |k\rangle_N \langle k'| \right. \right.$$
$$\left. \left. \cdot \sum_{ll',mm'} \mathbf{F}^{mm'}_{l'l} \left|{}^{l'}_{m}\right\rangle_{N^*\otimes N} \left\langle{}^{l}_{m'}\right| \right) \right)$$
$$= \sum_{ii',jj',ll'} \mathbf{T}^{ii'}_{j'j} \mathbf{P}^{ll'} \mathbf{F}^{jj'}_{l'l} |i\rangle_N \langle i'|. \qquad (55)$$

The respective graphical representations of these contractions can be found in fig.2. Though the coefficients might be different for each derivation, it is not clear how both interpretations are carried separately if they are part of a larger fragment, since their description takes place on the same space. Also, this recipe gives a fixed ordering and range for each trace. To be able to describe each final meaning separately, we use here the concept of *subsystem*. Because different subsystems act formally as different syntactic types and in each derivation the words that interact are different, it follows that each word should be assigned to a different subsystem:

- $[\![tall^{n/n}]\!]_{I_1} = [\![tall^{n/n}]\!]_{I_2} = \sum_{ii',jj'} \mathbf{T}^{ii'}_{j'j} \left|{}^{j'}_{i}\right\rangle_{N^1\otimes N^{2*}} \left\langle{}^{j}_{i'}\right|,$
- $[\![person^n]\!]_{I_1} = \sum_{kk'} \mathbf{P}^{kk'} |k\rangle_{N^2} \langle k'|,$
  $[\![person^n]\!]_{I_2} = \sum_{kk'} \mathbf{P}^{kk'} |k\rangle_{N^3} \langle k'|,$

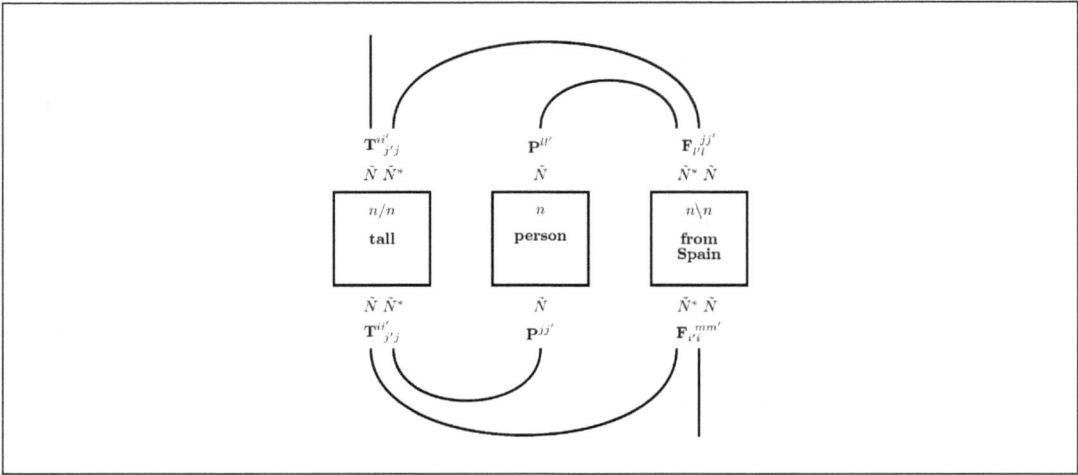

Figure 2: Representation of contractions corresponding to the first reading (lower links) and to the second reading (upper links), without subsystems. The final value is a coefficient in the $\tilde{N}$ space as in eq.54 and in eq.55, respectively.

- $[\![from\_Spain^{n\backslash n}]\!]_{I_1} = \sum_{ll',mm'} \mathbf{F}_{l'l}^{mm'} \left|{l' \atop m}\right\rangle_{N^{1*} \otimes N^3} \left\langle{l \atop m'}\right|,$
  $[\![from\_Spain^{n\backslash n}]\!]_{I_2} = \sum_{ll',mm'} \mathbf{F}_{l'l}^{mm'} \left|{l' \atop m}\right\rangle_{N^{3*} \otimes N^2} \left\langle{l \atop m'}\right|.$

Notice that the value of the coefficients given by the interpretation functions $I_1$ and $I_2$ that describe the words does not change from the ones given in $I$, only possibly the subsystem assignment does. Rewriting the derivation of the interpretations in terms of subsystems, the ordering of the traces does not matter anymore since the contraction is restricted to its own subsystem. For the first reading we obtain

$$[\![tall\_person\_from\_Spain^{n}]\!]_{I_1}^{1} =$$
$$= \mathrm{Tr}_{N^1}\left(\mathrm{Tr}_{N^2}\left(\sum_{ii',jj'} \mathbf{T}_{j'j}^{ii'} \left|{j' \atop i}\right\rangle_{N^1 \otimes N^{2*}} \left\langle{j \atop i'}\right| \cdot \sum_{kk'} \mathbf{P}^{kk'} |k\rangle_{N^2} \langle k'|\right.\right.$$
$$\left.\left.\cdot \sum_{ll',mm'} \mathbf{F}_{l'l}^{mm'} \left|{l' \atop m}\right\rangle_{N^{1*} \otimes N^3} \left\langle{l \atop m'}\right|\right)\right)$$
$$= \sum_{ii',jj',mm'} \mathbf{T}_{j'j}^{ii'} \mathbf{P}^{jj'} \mathbf{F}_{i'i}^{mm'} |m\rangle_{N^3} \langle m'| \qquad (56)$$

and for the second

$$[\![tall\_person\_from\_Spain^n]\!]^2_{I_2} =$$

$$= \text{Tr}_{N^2} \left( \sum_{ii',jj'} \mathbf{T}^{ii'}_{j'j} \left|^{j'}_i\right\rangle_{N^1 \otimes N^{2*}} \left\langle^j_{i'}\right| \cdot \text{Tr}_{N^3} \left( \sum_{kk'} \mathbf{P}^{kk'} |k\rangle_{N^3} \langle k'| \right. \right.$$

$$\left. \left. \cdot \sum_{mm',ll'} \mathbf{F}_{l'l}^{mm'} \left|^{l'}_m\right\rangle_{N^{3*} \otimes N^2} \left\langle^l_{m'}\right| \right) \right)$$

$$= \text{Tr}_{N^3} \left( \text{Tr}_{N^2} \left( \sum_{ii',jj'} \mathbf{T}^{ii'}_{j'j} \left|^{j'}_i\right\rangle_{N^1 \otimes N^{2*}} \left\langle^j_{i'}\right| \cdot \sum_{kk'} \mathbf{P}^{kk'} |k\rangle_{N^3} \langle k'| \right. \right.$$

$$\left. \left. \cdot \sum_{ll',mm'} \mathbf{F}_{l'l}^{mm'} \left|^{l'}_m\right\rangle_{N^{3*} \otimes N^2} \left\langle^l_{m'}\right| \right) \right)$$

$$= \sum_{ii',jj',ll'} \mathbf{T}^{ii'}_{j'j} \mathbf{P}^{ll'} \mathbf{F}_{l'l}^{jj'} |i\rangle_{N^1} \langle i'|. \tag{57}$$

The interpretation of each derivation belongs now to different subsystems, which keeps the information about the original word to which the free "noun" space is attached. We can see this by comparing the upper and lower links in fig.3.

However, it is not very convenient to attribute each word to a different subsystem depending on the interpretation it will be part of, since that is information that comes from the derivation itself and not from the representations of words. To tackle this problem, one uses permutation operations over the subsystems. Since these have precedence over the trace, when the traces are taken the contractions change accordingly. This changes the subsystem assignment at specific points so it is possible to go from one interpretation to the other, without giving different interpretations to each word initially. Thus, there is a way to go directly from the first interpretation to the second:

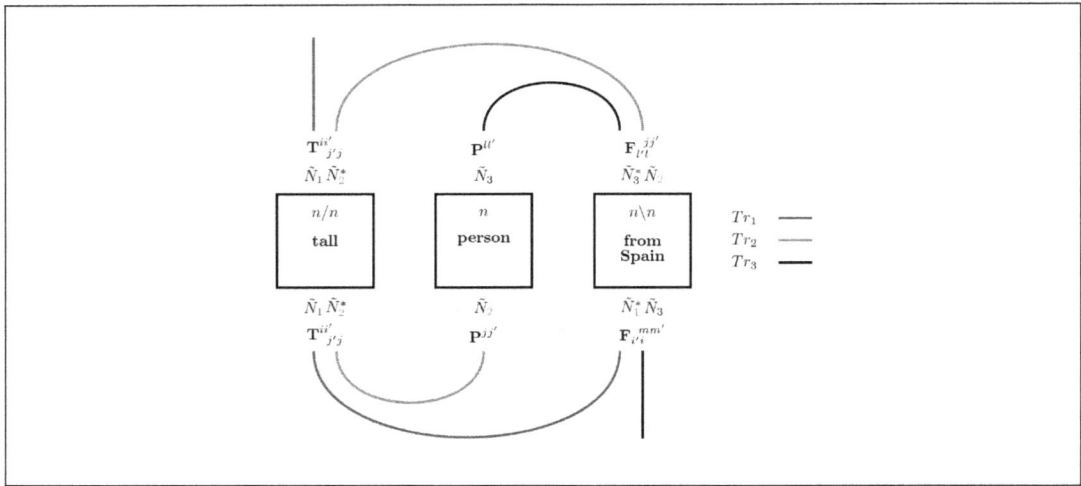

Figure 3: Representation of contractions corresponding to the first reading (lower links) and to the second reading (upper links), with subsystems. The final value is a coefficient in the $\tilde{N}$ space as in eq.56 and in eq.57, respectively.

$$[\![tall\_person\_from\_Spain^n]\!]^2_{I_1} =$$

$$= \text{Tr}_{N^1}\left(P_{13}\,\text{Tr}_{N^2}\left(\sum_{ii',jj'}\mathbf{T}^{ii'}_{j'j}\left|{j'\atop i}\right\rangle_{N^1\otimes N^{2*}}\left\langle{j\atop i'}\right| \cdot P_{13}P^{23}\sum_{kk'}\mathbf{P}^{kk'}\left|k\right\rangle_{N^2}\left\langle k'\right|\right.\right.$$

$$\left.\left.\cdot \sum_{ll',mm'}\mathbf{F}^{mm'}_{l'l}\left|{l'\atop m}\right\rangle_{N^{1*}\otimes N^3}\left\langle{l\atop m'}\right|P^{23}P_{13}\right)P_{13}\right)$$

$$= \text{Tr}_{N^3}\left(\text{Tr}_{N^2}\left(\sum_{ii',jj'}\mathbf{T}^{ii'}_{j'j}\left|{j'\atop i}\right\rangle_{N^1\otimes N^{2*}}\left\langle{j\atop i'}\right| \cdot \sum_{kk'}\mathbf{P}^{kk'}\left|k\right\rangle_{N^3}\left\langle k'\right|\right.\right.$$

$$\left.\left.\cdot \sum_{ll',mm'}\mathbf{F}^{mm'}_{l'l}\left|{l'\atop m}\right\rangle_{N^{3*}\otimes N^2}\left\langle{l\atop m'}\right|\right)\right). \qquad (58)$$

The reasoning behind this is as follows: the permutation $P^{23}$ swaps the space assignment between that of "person" and the free space in "from_Spain", according to eq.42; after that a permutation $P_{13}$ is used as in eq.39 to change the argument space of "from_Spain" from $N^{1*}$ to $N^{3*}$, and then the same permutation is applied again to change the space of tracing, following eq.40. In this way, all the coefficients

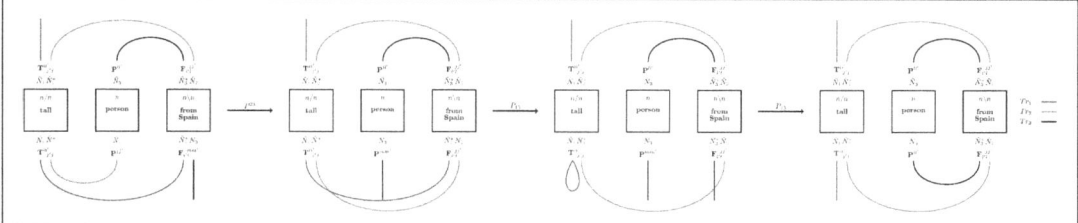

Figure 4

will have the correct contractions and in a different space from the first reading. In fig.4 we can see the action of the permutations by visualizing how both the spaces and the traces change as we go from the lower to the upper links.

Although the metric is not used explicitly in the application of the permutation operators, it is necessary to generate the correct tensors where the permutation operator is applied in the first place, by going from the vector representation that comes directly from the data to one that allows contraction. As an example, the adjective "tall" would have a vector representation from the data as an element of $\tilde{V} \otimes \tilde{V}$, of the form $\mathbf{T}^{ii',kk'}$. We need the metric $d_{kj'}d_{k'j}$ to change its form to $T^{ii'}_{j'j}$. By defining the interpretation space of adjectives as $\tilde{N} \otimes \tilde{N}^*$, we assume this passage has already been made when we assign an interpretation to a word in this space. As an alternative to this derivation, we mention that it is possible to apply a $P^{23}$ permutation followed by a $P^{13}$ permutation that results in the correct contraction of the indices, but fails to deliver the results of the two derivations in different subspaces; it is however noteworthy that, in order to start with a unique assignment for each word, this alternative derivation can, in any case, only be achieved by distinguishing between subsystems, as well as the covariant and contravariant indices.

## 7 Conclusion and Future Work

In this paper we provided a density matrix model for a simple fragment of the Lambek Calculus, differently from what is done in [2] who uses density matrices to interpret dependency parse trees. The syntax-semantics interface takes the form of a compositional map assigning semantic values to the $\lambda_{/,\backslash}$ terms coding syntactic derivations. We proposed the use of a metric as a way to reconcile the various vector representations of the same word that come from different treatments, assuming that there is a quantity that is being preserved, such as human judgements. If we know the metric, we can confidently assign only one embedding to each word as its semantic value. A metric can relate these various representations so that we can assign only one vector as its semantic value. The density matrix model enables the integration of lexical and derivational forms of ambiguity. Additionally, it allows for the transfer of methods and techniques from quantum mechanics and general relativity to computational semantics. One example of such transfer is the permutation operator. In quantum mechanics, this operator permits a description of indistinguishable particles. In the linguistic application, it allows one to go from an interpretation that comes from one derivation to another, without the need to to go through the latter, but keeping this second meaning in a different subsystem. Another example is the introduction of covariant and contravariant components, associated with a metric, that allow the permutation operations to be properly applied. In future work, we want to explore the preservation of human judgements found in the literature via a metric that represents the variability of vector representations of words, either static or dynamic. We also want to extend our simple fragment with modalities for structural control (cf [17]), in order to deal with cases of derivational ambiguity that are licensed by these control modalities. Finally, we want to consider derivational ambiguity in the light of an *incremental* left-to-right interpretation process, so as to account for the evolution of interpretations over time. In enriching the treatment with a metric, we want to explore the consequences of having this new parameter in treating context dependent embeddings.

# Acknowledgements

A.D.C. thanks discussions with Sanjaye Ramgoolam and Martha Lewis that contributed to the duality concepts included in the journal version of this paper. This work is supported by the Complex Systems Fund, with special thanks to Peter Koeze.

# References

[1] Dea Bankova, Bob Coecke, Martha Lewis, and Dan Marsden. Graded hyponymy for compositional distributional semantics. *Journal of Language Modelling*, 6(2):225–260, 2019.

[2] William Blacoe. Semantic composition inspired by quantum measurement. In *International Symposium on Quantum Interaction*, pages 41–53. Springer, 2014.

[3] Elia Bruni, Nam-Khanh Tran, and Marco Baroni. Multimodal distributional semantics. *Journal of Artificial Intelligence Research*, 49:1–47, 2014.

[4] Stephen Clark. Vector space models of lexical meaning. *Handbook of Contemporary Semantics*, 10:9781118882139, 2015.

[5] Bob Coecke, Mehrnoosh Sadrzadeh, and Stephen Clark. Mathematical foundations for a compositional distributional model of meaning. *Lambek Festschrift, Linguistic Analysis 36(1-4)*, pages 345–384, 2010.

[6] AD Correia, HTC Stoof, and M Moortgat. Putting a spin on language: A quantum interpretation of unary connectives for linguistic applications. *arXiv preprint arXiv:2004.04128*, 2020.

[7] Jacob Devlin, Ming-Wei Chang, Kenton Lee, and Kristina Toutanova. BERT: Pre-training of deep bidirectional transformers for language understanding. In *Proceedings of the 2019 Conference of the North American Chapter of the Association for Computational Linguistics: Human Language Technologies, Volume 1 (Long and Short Papers)*, pages 4171–4186, Minneapolis, Minnesota, June 2019. Association for Computational Linguistics.

[8] Kees Dullemond and Kasper Peeters. Introduction to tensor calculus. *Kees Dullemond and Kasper Peeters*, 1991.

[9] Howard Georgi. Lie algebras in particle physics. from isospin to unified theories. *Front. Phys.*, 54:1–255, 1982.

[10] Edward Grefenstette and Mehrnoosh Sadrzadeh. Experimental support for a categorical compositional distributional model of meaning. In *Proceedings of the Conference on Empirical Methods in Natural Language Processing*, pages 1394–1404. Association for Computational Linguistics, 2011.

[11] Zellig S Harris. Distributional structure. *Word*, 10(2-3):146–162, 1954.

[12] Theo M.V. Janssen and Barbara H. Partee. Chapter 7 - Compositionality. In Johan van Benthem and Alice ter Meulen, editors, *Handbook of Logic and Language*, pages 417 – 473. North-Holland, Amsterdam, 1997.

[13] Joachim Lambek. The mathematics of sentence structure. *The American Mathematical Monthly*, 65(3):154–170, 1958.

[14] Joachim Lambek. On the calculus of syntactic types. In Roman Jakobson, editor, *Structure of Language and its Mathematical Aspects*, volume XII of *Proceedings of Symposia in Applied Mathematics*, pages 166–178. American Mathematical Society, 1961.

[15] Tomas Mikolov, Ilya Sutskever, Kai Chen, Greg S Corrado, and Jeff Dean. Distributed representations of words and phrases and their compositionality. In *Advances in neural information processing systems*, pages 3111–3119, 2013.

[16] Jeff Mitchell and Mirella Lapata. Composition in distributional models of semantics. *Cognitive science*, 34(8):1388–1429, 2010.

[17] Michael Moortgat. Chapter 2 - Categorial type logics. In Johan van Benthem and Alice ter Meulen, editors, *Handbook of Logic and Language*, pages 93–177. Elsevier, Amsterdam, 1997.

[18] Richard Moot and Christian Retoré. *The logic of categorial grammars: a deductive account of natural language syntax and semantics*, volume 6850. Springer, 2012.

[19] Michael A Nielsen and Isaac Chuang. Quantum computation and quantum information, 2002.

[20] Jeffrey Pennington, Richard Socher, and Christopher D Manning. Glove: Global vectors for word representation. In *Proceedings of the 2014 conference on empirical methods in natural language processing (EMNLP)*, pages 1532–1543, 2014.

[21] Matthew Peters, Mark Neumann, Mohit Iyyer, Matt Gardner, Christopher Clark, Kenton Lee, and Luke Zettlemoyer. Deep contextualized word representations. In *Proceedings of the 2018 Conference of the North American Chapter of the Association for Computational Linguistics: Human Language Technologies, Volume 1 (Long Papers)*, pages 2227–2237, New Orleans, Louisiana, June 2018. Association for Computational Linguistics.

[22] Robin Piedeleu. *Ambiguity in categorical models of meaning*. PhD thesis, University of Oxford Master's thesis, 2014.

[23] Robin Piedeleu, Dimitri Kartsaklis, Bob Coecke, and Mehrnoosh Sadrzadeh. Open system categorical quantum semantics in natural language processing. *CoRR*, abs/1502.00831, 2015.

[24] Mehrnoosh Sadrzadeh, Dimitri Kartsaklis, and Esma Balkir. Sentence entailment in compositional distributional semantics. *Ann. Math. Artif. Intell.*, 82(4):189–218, 2018.

[25] Lorenzo Adlai Sadun. *Applied linear algebra: The decoupling principle*. American Mathematical Soc., 2007.

[26] Peter Selinger. Dagger compact closed categories and completely positive maps. *Electronic Notes in Theoretical computer science*, 170:139–163, 2007.

[27] Johan van Benthem. The semantics of variety in categorial grammar. Technical Report 83-29, Simon Fraser University, Burnaby (B.C.), 1983. Revised version in W. Buszkowski, W. Marciszewski and J. van Benthem (eds) Categorial grammar, Benjamin, Amsterdam.

[28] Robert M Wald. General relativity. *University of Chicago Press*, 1984.

[29] Heinrich Wansing. Formulas-as-types for a hierarchy of sublogics of intuitionistic propositional logic. In David Pearce and Heinrich Wansing, editors, *Nonclassical Logics and Information Processing*, pages 125–145, Berlin, Heidelberg, 1992. Springer Berlin Heidelberg.

[30] Gijs Wijnholds and Mehrnoosh Sadrzadeh. Evaluating composition models for verb phrase elliptical sentence embeddings. In *Proceedings of the 2019 Conference of the North American Chapter of the Association for Computational Linguistics: Human Language Technologies, Volume 1 (Long and Short Papers)*, pages 261–271, 2019.

# A Frobenius Algebraic Analysis for Parasitic Gaps

Michael Moortgat
*Utrecht University**
m.j.moortgat@uu.nl

Mehrnoosh Sadrzadeh
*University College London*[†]
m.sadrzadeh@ucl.ac.uk

Gijs Wijnholds
*Utrecht University*
g.j.wijnholds@uu.nl

## Abstract

The interpretation of parasitic gaps is an ostensible case of non-linearity in natural language composition. Existing categorial analyses, both in the typelogical and in the combinatory traditions, rely on explicit forms of syntactic copying. We identify two types of parasitic gapping where the duplication of semantic content can be confined to the lexicon. Parasitic gaps in *adjuncts* are analysed as forms of generalized coordination with a polymorphic type schema for the head of the adjunct phrase. For parasitic gaps affecting *arguments* of the same predicate, the polymorphism is associated with the lexical item that introduces the primary gap. Our analysis is formulated in terms of Lambek calculus extended with structural control modalities. A compositional translation relates syntactic types and derivations to the interpreting compact closed category of finite dimensional vector spaces and linear maps with Frobenius algebras over it. When interpreted over the necessary semantic spaces, the Frobenius algebras provide the tools to model the proposed instances of lexical polymorphism.

---

[*]The research of the alphabetically first and third author is supported by NWO grant 360-89-070 "A composition calculus for vector-based semantic modelling with a localization for Dutch".

[†]The alphabetically second author acknowledges the support of Royal Society International Exchange Award IE161631.

# 1 Introduction

Natural languages present many situations where an overt syntactic element provides the semantic content for one or more occurrences of elements that are not physically realized, or that have no meaning of their own. Illustrative cases can be found at the sentence and at the discourse level, e.g. long-distance dependencies in 'movement' constructions, ellipsis phenomena, anaphora. Parasitic gaps are a challenging case in point.

To provide the reader with the necessary linguistic background, the examples in (1) illustrate some relevant patterns[1]. The symbol ␣ marks the position of the virtual elements that depend on a physically realized phrase elsewhere in their context for their interpretation; in the generative grammar literature, these virtual elements are referred to as "gaps".

$$
\begin{aligned}
&a \quad \text{papers that Bob rejected ␣ (immediately)} \\
&b \quad \text{Bob left the room without closing the window} \\
&c \quad \text{*window that Bob left the room without closing ␣} \\
&d \quad \text{papers that Bob rejected ␣ without reading ␣}_p \text{ (carefully)} \\
&e \quad \text{security breach that a report about ␣}_p \text{ in the NYT made ␣ public} \\
&f \quad \text{this is a candidate whom I would persuade every friend of ␣ to vote for ␣}
\end{aligned}
\quad (1)
$$

Consider first the case of object relativisation in (a). This example has a single gap for the unexpressed direct object of *rejected*. In categorial type logics, gaps have the status of *hypotheses*, introduced by a higher-order type. In Lambek's [6] Syntactic Calculus, for example, the relative pronoun *that* in (a) would be typed as $(n\backslash n)/(s/np)$. The complete relative clause then acts as a noun postmodifier $n\backslash n$. The relative clause body *Bob rejected* ␣ is typed as $s/np$, which means it needs a noun phrase hypothesis in order to compose a full sentence. Because the hypothesis occupies the direct object position, it is impossible to physically realize that object, as the ungrammaticality of *papers that Bob rejected the proposal* shows. The Lambek type requires the hypothetical $np$ to occur at the right periphery of the relative clause body — a restriction that we will lift in Section §2 to allow for phrase-internal hypotheses. An example would be (a) with an extra temporal adverb (*immediately*) at the end.

As the name suggests, a *parasitic* gap is felicitous only in the presence of a primary gap. The relative clause in (d) has two gaps: the primary one is for the

---

[1]For a more thorough discussion of the phenomena, and proposed analyses in a variety of grammatical frameworks, see[2].

object of *rejected* as in (a); the secondary, parasitic gap (marked by $\sqcup_p$) is the unexpressed object of *reading*. The parasitic gap occurs here in an adjunct: the verb phrase modifier *without closing* $\sqcup$. Such an adjunct by itself, is an *island* for extraction: the ungrammatical (c) shows that it is impossible for the relative pronoun to establish communication with a *np* hypothesis occuring within the adjunct phrase. Compare (c) with the gapless (b) which has the complete adjunct *without closing (the window)$_{np}$*.

Examples (e) and (f) represent a different type of parasitic gapping where both the primary and the parasitic gap regard co-arguments of the same verb. In (e), the primary gap is the direct object of *made public*, the secondary gap occurs in the subject argument of this predicate. In (f), the primary gap is the object of the infinitive complement of the verb *persuade*, viz. *to vote for* $\sqcup$, while the secondary gap occurs in the direct object of *persuade*[2].

We illustrated the adjunct and co-argument types of parasitic gapping in (1) with relative clause examples. Primary gaps can also be triggered in main or subordinate constituent question constructions, as in (2a, b), where *which papers* will carry the higher-order type initiating hypothetical reasoning. In the 'passive infinitive' case (2c), the higher-order type is associated with the adjective *hard*, which in this context could be typed as $ap/(to\_inf/np)$. The adjective then selects for an incomplete *to*-infinitive missing a *np* hypothesis, the direct object in (2c). As with the relative clause example (1a), putting a physically realized *np* in the position occupied by the hypothesis leads to ungrammaticality. Again, as in (1), the primary gaps here open the possibility for parasitic gaps dependent on them as in (2d, e, f). These examples also illustrate some of the various forms the adjunct phrase can take: temporal modification (*before, after*), contrastive (*despite*), etc.

    a  which papers did Bob reject $\sqcup$ (immediately)
    b  I know which papers Bob will reject $\sqcup$ (immediately)
    c  this paper is hard to understand $\sqcup$ / *the proposal
    d  which papers did Bob accept $\sqcup$ *despite* not liking $\sqcup_p$ (really)
    e  I know which papers Bob will reject $\sqcup$ *before* even reading $\sqcup_p$ (cursorily)
    f  this paper is easy to explain $\sqcup$ well *after* studying $\sqcup_p$ (thoroughly)

(2)

To account for the duplication of semantic content in parasitic gap constructions, existing categorial analyses rely on explicit forms of syntactic copying. The CCG analysis of [19] rests on (a directional version of) the **S** combinator of Combinatory Logic; the type-logical account of [13, 14] adapts the ! modality of Linear Logic to

---

[2] According to [19], each of the gaps in this type of example would be felicitous by itself.

implement a restricted form of the structural rule of Contraction. These syntactic devices are hard to control: the CCG version of the **S** rule is constrained by rule features; the attempts to properly constrain Contraction easily lead to undecidabilty as shown in [4].

Our aim in this paper is to explore *lexical polymorphism* as an alternative to syntactic copying. The technique of polymorphic typing is standardly used in categorial grammars for chameleon words such as *and, but*. Rather than giving these words a single type, they are assigned a type *schema*, with different realizations depending on whether they are conjoining sentences, verb phrases, transitive verbs, etc. Treating the adjunct phrases of $(1d)$ and $(2d, e, f)$ as forms of *subordinating* conjunction, we propose to similarly handle the adjunct type of parasitic gaps by means of a polymorphic type schema for the heads *without, despite, after*, etc. In the co-argument type of parasitic gapping $(1e, f)$, a conjunctive interpretation is absent. In this case, a polymorphic type schema for the relative pronouns *that* or *who(m)* allows us to generalize from the single gap instance $(1a)$ to the multi-gap case $(1e)$. To obtain the derived relative pronoun type from the basic assignment, we can rely on the same mechanisms that relate the basic type for *without* etc to the derived type needed for the parasitic gap examples.

Our analysis builds on the categorial Frobenius algebraic compositional distributional semantics of [16, 17], combined with a multimodal extension of Lambek calculus as the syntactic front end, as in [9]. Our analysis provides further evidence that Frobenius algebra is a powerful tool to model the internal dynamics of lexical semantics.

## 2 Syntax

### 2.1 The logic $\mathbf{NL}_\diamond$

The syntactic front end for our analysis is the type logic $\mathbf{NL}_\diamond$ of [10] which extends Lambek's pure logic of residuation [7] with modalities for structural control. The formula language is given by the following grammar ($p$ atomic):

$$A, B ::= p \mid A \otimes B \mid A/B \mid A\backslash B \mid \Diamond A \mid \Box A \tag{3}$$

In $\mathbf{NL}_\diamond$, types are assigned to *phrases*, not to strings as in the more familiar Syntactic Calculus of [6], or its pregroup version [8]. The tensor product $\otimes$ then is a non-associative, non-commutative operation for putting phrases together; it has adjoints $/$ and $\backslash$ expressing right and left incompleteness with respect to phrasal composition, as captured by the residuation inferences (4). In addition to the binary family $/, \otimes, \backslash$, the extended language has unary control modalities $\Diamond, \Box$ which again form a residuated pair with the inferences in (5).

$$A \longrightarrow C/B \quad \text{iff} \quad A \otimes B \longrightarrow C \quad \text{iff} \quad B \longrightarrow A\backslash C \qquad (4)$$

$$\Diamond A \longrightarrow B \quad \text{iff} \quad A \longrightarrow \Box B \qquad (5)$$

The modalities serve a double purpose, either *licensing* reordering or restructuring that would otherwise be forbidden, or *blocking* structural operations that otherwise would be applicable. To license rightward extraction, as found in English long-range dependencies, we use the postulates in (6). Postulate $\alpha_\diamond$ is a controlled form of associativity: the $\Diamond$ marking licenses a rotation of the tensor formula tree that leaves the order of the components $A, B, \Diamond C$ unaffected. Postulate $\sigma_\diamond$ implements a form of controlled commutativity: here the internal structure of the tensor formula tree is unaffected, but the components $B$ and $\Diamond C$ are exchanged.

$$\begin{aligned} \alpha_\diamond &: (A \otimes B) \otimes \Diamond C \longrightarrow A \otimes (B \otimes \Diamond C) \\ \sigma_\diamond &: (A \otimes B) \otimes \Diamond C \longrightarrow (A \otimes \Diamond C) \otimes B \end{aligned} \qquad (6)$$

To block these structural operations from applying, we use a pair of modalities $\Diamond, \Box$. Phrases that qualify as syntactic islands are marked off by $\Diamond$. The modal island demarcation makes sure that the input conditions for $\alpha_\diamond, \sigma_\diamond$ do not arise. The island markers $\Diamond, \Box$ have no associated structural rules; their logical behaviour is fully characterized by (5).

**NL**$_\diamond$ derivations will be represented using the axiomatisation of Figure 1, due to Došen [3]. This axiomatisation takes (Co)Evaluation as primitive arrows, and recursively generalizes these by means of Monotonicity. It is routine to show that the residuation inferences of (4) and (5) become derivable rules given the axiomatisation of Figure 1. To streamline derivations, we will make use of the derived residuation steps. Also, we will freely use (Co)Evaluation and the structural postulates (6) in their *rule* form, by composing them with Transitivity ($\circ$).

## 2.2 Graphical calculus for NL$_\diamond$

Wijnholds [23] gives a coherent diagrammatic language for the non-associative Lambek Calculus **NL**; the generalisation to **NL** with control modalities is straightforward, see Figure 2. In short, each connective is assigned two *links* that either compose or decompose a type built with that connective. Links (and diagrams) can be put together granted that their in- and outputs coincide. This system has a full recursive definition, and is shown to be sound and complete (i.e. coherent) with respect to the categorical formulation of the Lambek Calculus, given a suitable set of graphical equalities (not discussed in the current paper).

$$1_A : A \longrightarrow A \qquad \frac{f : A \longrightarrow B \quad g : B \longrightarrow C}{g \circ f : A \longrightarrow C}$$

$$\frac{f : A \longrightarrow B \quad g : C \longrightarrow D}{f \otimes g : A \otimes C \longrightarrow B \otimes D}$$

$$\frac{f : A \longrightarrow B \quad g : C \longrightarrow D}{f/g : A/D \longrightarrow B/C} \qquad \frac{f : A \longrightarrow B \quad g : C \longrightarrow D}{f \backslash g : B \backslash C \longrightarrow A \backslash D}$$

$$\frac{f : A \longrightarrow B}{\Diamond f : \Diamond A \longrightarrow \Diamond B} \qquad \frac{f : A \longrightarrow B}{\Box f : \Box A \longrightarrow \Box B}$$

$$ev^{\backslash}_{A,B} : A \otimes A\backslash B \longrightarrow B \qquad \text{co-}ev^{\backslash}_{A,B} : B \longrightarrow A\backslash(A \otimes B)$$

$$ev^{/}_{A,B} : B/A \otimes A \longrightarrow B \qquad \text{co-}ev^{/}_{A,B} : B \longrightarrow (B \otimes A)/A$$

$$ev^{\Box}_A : \Diamond\Box A \longrightarrow A \qquad \text{co-}ev^{\Box}_A : A \longrightarrow \Box\Diamond A$$

$$\alpha_\Diamond : (A \otimes B) \otimes \Diamond C \longrightarrow A \otimes (B \otimes \Diamond C) \qquad \sigma_\Diamond : (A \otimes B) \otimes \Diamond C \longrightarrow (A \otimes \Diamond C) \otimes B$$

Figure 1: Došen style axiomatisation of **NL**$_\Diamond$.

As an illustration, we present the derivation of the simple relative clause example (1a) in symbolic and diagrammatic form. For this case of non-subject[3] relativisation, the relative pronoun *that* is typed as a functor that produces a noun modifier $n\backslash n$ in combination with a sentence that contains an unexpressed $np$ hypothesis (*Bob rejected ⊔ immediately*). The subtype for the gap is the modally decorated formula $\Diamond\Box np$. The $\Diamond$ marking allows it to cross phrase boundaries on its way to the phrase-internal position adjacent to the transitive verb *rejected*. At that point, the licensing $\Diamond$ has done its work, and can be disposed of by means of the $ev^\Box$ axiom $\Diamond\Box np \longrightarrow np$, which provides the $np$ object required by the transitive verb *rejected*. For legibility, we use words instead of their types for the lexical assumptions in the derivation below. The steps labeled $\ell$ indicate the lexical look-up.

---

[3]Subject relative clauses, e.g. *paper that ⊔ irritates Bob*, do not involve any structural reasoning. The relative pronoun for subject relatives can be typed simply as $(n\backslash n)/(np\backslash s)$.

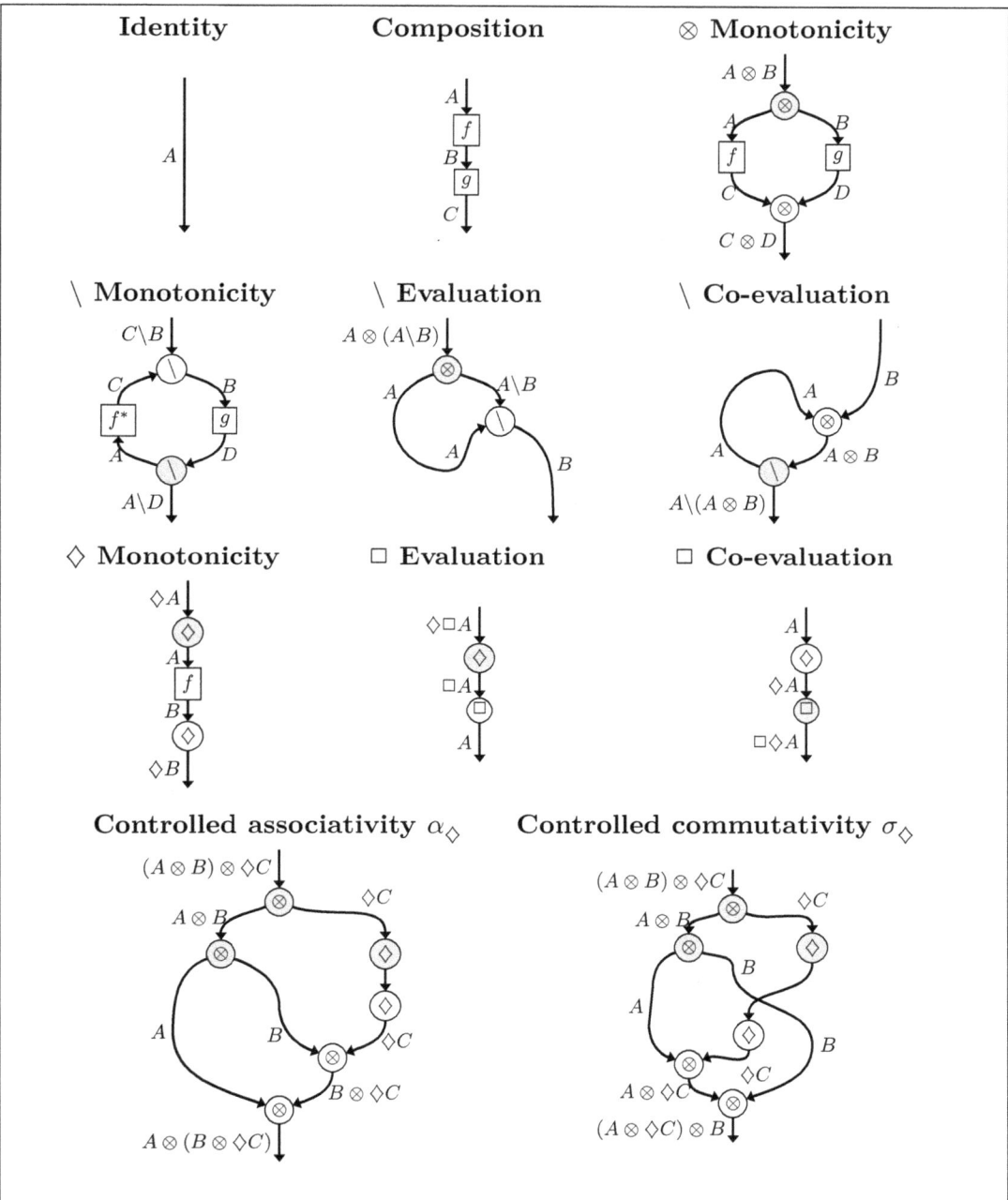

Figure 2: Došen style axiomisation of **NL**$_\Diamond$ with diagrams. Monotonicity and (co)evaluation laws for / are fully symmetrical to the given diagrams for \.

$$\dfrac{\text{paper}}{n}\,\ell\quad \dfrac{\dfrac{\text{that}}{(n\backslash n)/(s/\Diamond\Box np)}\,\ell\quad \dfrac{\dfrac{\text{Bob}}{np}\,\ell\quad \dfrac{\dfrac{\dfrac{\text{rejected}}{(np\backslash s)/np}\,\ell\quad \overline{\Diamond\Box np \longrightarrow np}\,ev^{\Box}}{\text{rejected} \otimes \Diamond\Box np \longrightarrow np\backslash s}\,ev^{/}\quad \dfrac{\text{immediately}}{(np\backslash s)\backslash(np\backslash s)}\,\ell}{(\text{rejected} \otimes \Diamond\Box np) \otimes \text{immediately} \longrightarrow np\backslash s}\,ev^{\backslash}}{\dfrac{\dfrac{\dfrac{\dfrac{\text{Bob} \otimes ((\text{rejected} \otimes \Diamond\Box np) \otimes \text{immediately}) \longrightarrow s}{\text{Bob} \otimes ((\text{rejected} \otimes \text{immediately}) \otimes \Diamond\Box np) \longrightarrow s}\,\sigma_{\diamond}}{(\text{Bob} \otimes (\text{rejected} \otimes \text{immediately})) \otimes \Diamond\Box np \longrightarrow s}\,\alpha_{\diamond}}{\text{Bob} \otimes (\text{rejected} \otimes \text{immediately}) \longrightarrow s/\Diamond\Box np}\,res_{/}}{\text{that} \otimes (\text{Bob} \otimes (\text{rejected} \otimes \text{immediately})) \longrightarrow n\backslash n}\,ev^{/}}\,ev^{\backslash}}{\text{paper} \otimes (\text{that} \otimes (\text{Bob} \otimes (\text{rejected} \otimes \text{immediately}))) \longrightarrow n}\,ev^{\backslash}$$

(7)

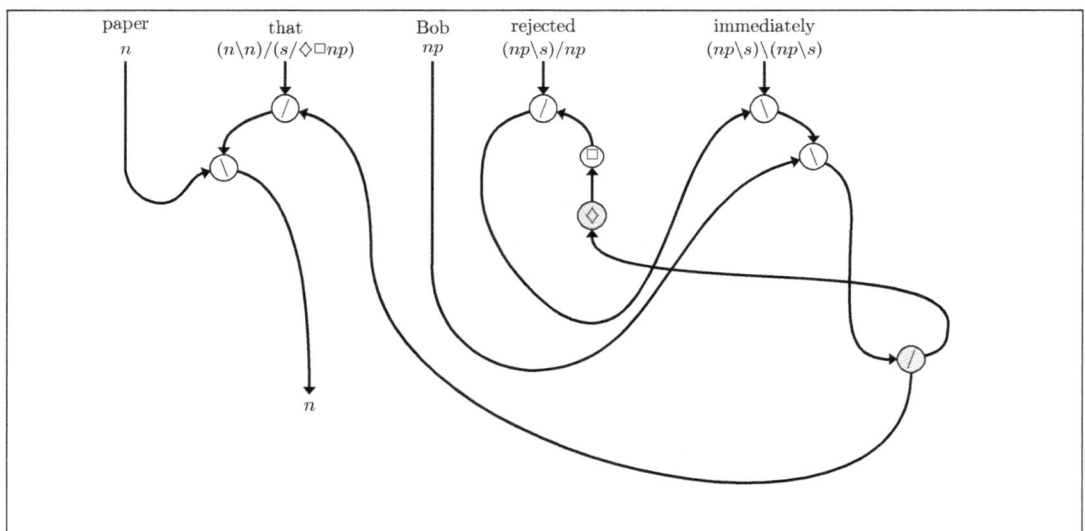

Figure 3: Diagrammatic form of *Paper that Bob rejected immediately*.

In the diagrammatic form of Fig 3, the $\Diamond\Box np$ gap hypothesis is indicated by the corresponding links. The leading $\Diamond$ link licenses the crossing over to the object position of *rejected* by means of the $\sigma_{\diamond}$ postulate of Fig 2. In what follows, we use diagrams for $\mathbf{NL}_{\diamond}$ derivations because this format pictures the information flow in a simple and intuitive way.

## 2.3 Typing Parasitic Gaps

**Lexical polymorphism: generalized coordination**  As our account of parasitic gaps in adjuncts treats the adjuncts as a form of subordinate conjunction, we briefly review how lexical polymorphism is used in the analysis of generalized coordination.

Chameleon words such as *and, but* cannot easily be typed monomorphically; given an initial type and interpretation, say $(s\backslash s)/s$ for sentence coordination, we'd like to be able to obtain derived types and interpretations for the coordination of (in)transitive verbs, as in $(8b, c)$, or for non-constituent coordination cases such as $(8d)$.

$$
\begin{aligned}
&a \quad \text{(Alice sings)}_s \text{ and (Bob dances)}_s \\
&b \quad \text{Alice (sings and dances)}_{np\backslash s} \\
&c \quad \text{Bob (criticized and rejected)}_{(np\backslash s)/np} \text{ the paper} \\
&d \quad \text{(Alice praised)}_{s/\Diamond\Box np} \text{ but (Bob criticized)}_{s/\Diamond\Box np} \text{ the paper}
\end{aligned}
\tag{8}
$$

Deriving the (b–d) types from an initial $(s\backslash s)/s$ assignment, however, goes beyond linearity. The attempt in (9) to derive verb phrase coordination from sentence coordination requires a copying step to strongly distribute the final $np$ abstraction over the two conjuncts.

$$
\cfrac{\cfrac{\cfrac{\vdots}{(\boxed{np}\otimes np\backslash s)\otimes((s\backslash s)/s\otimes(\boxed{np}\otimes np\backslash s))\longrightarrow s}}{\boxed{np}\otimes(np\backslash s\otimes((s\backslash s)/s\otimes np\backslash s))\longrightarrow s}}{(s\backslash s)/s\longrightarrow((np\backslash s)\backslash(np\backslash s))/(\boxed{np}\backslash s)} \text{ Copy!}
\tag{9}
$$

Partee and Rooth's [15] work on generalized coordination offers a method for replacing syntactic copying by lexical polymorphism. Coordinating expressions *and, but* get a polymorphic type assignment $(X\backslash X)/X$ where $X$ is a conjoinable type. The set of conjoinable types CType forms a subset of the general set of types Type. CType is defined inductively[4]:

- $s \in \text{CType}$;
- $A\backslash B, B/A \in \text{CType}$ if $B \in \text{CType}, A \in \text{Type}$

The type polymorphism comes with a generalized interpretation. We write $\sqcap^X$ (infix notation) for a coordinator of (semantic) type $X \to X \to X$.

---

[4] Partee and Rooth formulate this in terms of the semantic types obtained from the syntax-semantics homomorphism $h$, with $h(s) = t$ (the type of truth values), $h(np) = e$ (individuals) and $h(A\backslash B) = h(B/A) = h(A) \to h(B)$.

- $P \sqcap^t Q := P \wedge Q$  coordination in type $t$ amounts to boolean conjunction
- $P \sqcap^{A \to B} Q := \lambda x^A.(P\ x) \sqcap^B (Q\ x)$  distributing the $x^A$ parameter over the conjuncts

The generalized interpretation scheme, then, associates a type transition such as (9) with the Curry-Howard program that would be associated with a derivation involving the copying step. In Section §3, we will obtain the same effect using the Frobenius algebras over our vector-based interpretations.

**Parasitic gaps in adjuncts**  Consider the type lexicon for the data in (1a–d)[5].

$$\begin{array}{rcl}
\text{papers, window} & :: & n \\
\text{that} & :: & (n\backslash n)/(s/\Diamond\Box np) \\
\text{Bob} & :: & np \\
\text{rejected} & :: & (np\backslash s)/np \\
\text{reading, closing} & :: & gp/np \\
\text{immediately, carefully} & :: & iv\backslash iv \\
\text{without} & :: & \Box(X\backslash Y)/Z \quad \text{(schematic)} \\
\text{without}^{b,c} & :: & \Box(iv\backslash iv)/gp \\
\text{without}^{d} & :: & \Box((iv/\Diamond\Box np)\backslash(iv/np))/(gp/\Diamond\Box np)
\end{array} \quad (10)$$

The gap-less example (1b) provides the motivation for the basic type assignment to *without* as a functor combining with a non-finite gerund clause $gp$ to produce a verb-phrase modifier $iv\backslash iv$. To impose island constraints, we use a pair of modalities $\Diamond, \Box$. In order to block the ungrammatical (1c), we follow [11] and lock the $iv\backslash iv$ result type with $\Box$; the matching $\Diamond$ needed to unlock it has the effect of demarcating the modifier phrase *without closing the window* as an island, represented in the diagram below by means of a dotted line.

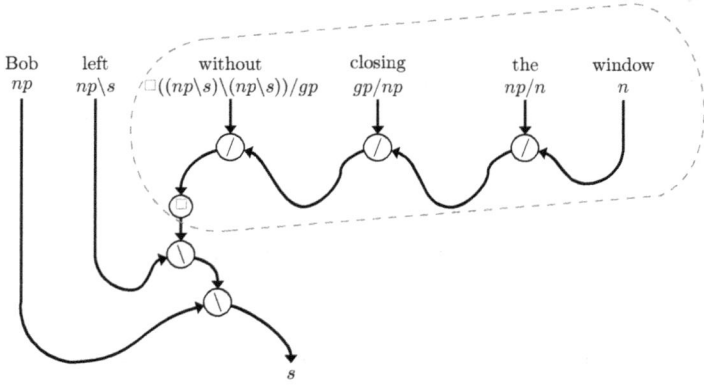

[5]$iv$ abbreviates $np\backslash s$; $gp$ stands for gerund clause, headed by the -ing form of the verb.

An attempt to derive the ungrammatical *window that Bob left without closing* ␣ fails. The derivation proceeds like the one above, but with the gap hypothesis $\Diamond\Box np$ in the place of *the window*. At that point the $\Diamond$ island demarcation of *without closing* $\Diamond\Box np$ makes it impossible to bring out the hypothesis to the position where it can be withdrawn. This becomes apparent diagrammatically as the gap hypothesis cannot cross the dotted line:

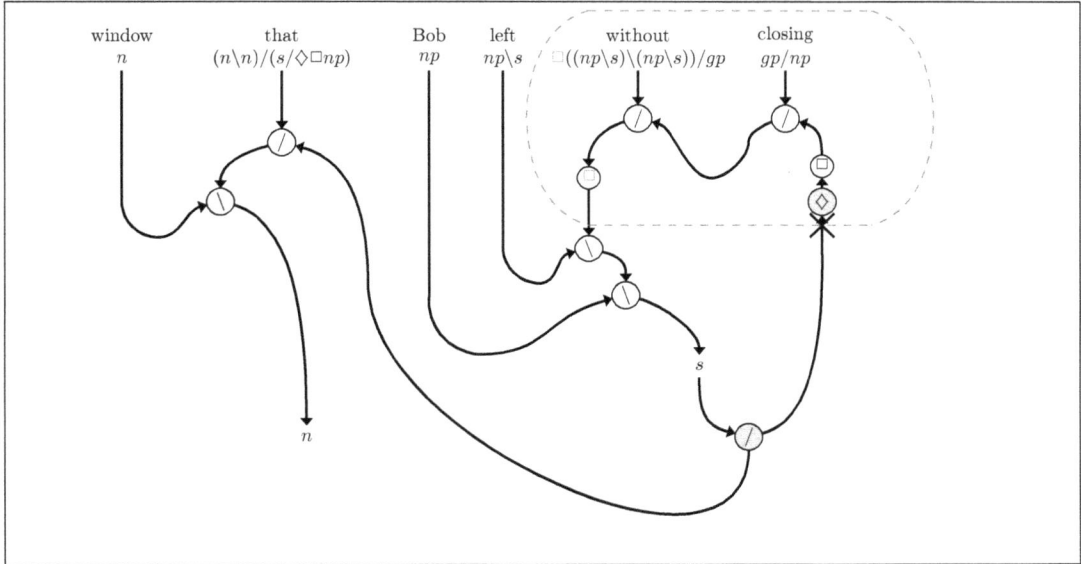

Let us turn then to the adjunct parasitic gapping of (1d). To account for the double use of the gap we replace *syntactic* copying via controlled Contraction by *lexical* polymorphism, treating *without* as a polymorphic item on a par with coordinators *and, but*. That means we assign to *without* the following type schema

$$\text{without} :: \Box(X\backslash Y)/Z$$

with basic instantiation $X = Y = iv$, $Z = gp$. From this basic instantiation, a derived instantiation with $X = Y = iv/\Diamond\Box np$ and $Z = gp/\Diamond\Box np$ is obtained for the parasitic gapping example (1d) by uniformly dividing the subtypes $iv$ and $gp$ by $\Diamond\Box np$ using the forward slash.

In Section §3, we will see how the vector-based interpretation of the derived type is obtained in a systematic fashion from the interpretation of the basic type instantiation. For this, it is helpful to factorize the construction of the derived type as the combination of an expansion step and a distribution step. Ignoring the appropriate $\Box$ decoration to mark off the adjunct as an island, the expansion step here is an instance of the Geach transformation $A/B \longrightarrow (A/C)/(B/C)$, with $A = iv\backslash iv$, $B = gp$, $C = \Diamond\Box np$.

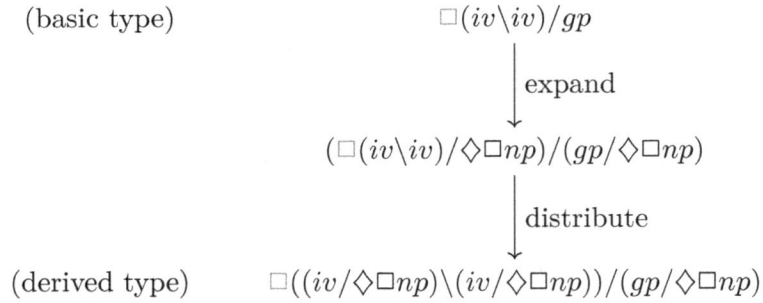

Setting now $A = iv$, $B = iv$, $C = \Diamond\Box np$, the distribution step is a directional instance of the **S** combinator $(A\backslash B)/C \longrightarrow (A/C)\backslash (B/C)$.

To arrive at the version of the derived type for *without* as we have it in our lexicon (10), a final calibration is required. We replace the result type $iv/\Diamond\Box np$ by $iv/np$, dropping the modal marking required for controlled associativity/commutativity. The final type $\Box((iv/\Diamond\Box np)\backslash (iv/np))/(gp/\Diamond\Box np)$ allows for the derivation of the parasitic gapping example (1c) displayed in Figure 4, but also for cases of Right Node Raising such as

Bob (rejected without reading)$_{iv/np}$ all papers about linguistics

where *all papers about linguistics* is a plain $np$ rather than $\Diamond\Box np$.

**Parasitic gaps: co-arguments**

Let us turn to the co-argument type of parasitic gapping as exemplified by (1e, f). Consider first (1e), repeated here for convenience, together with a gap-less sentence that motivates the type-assignments given in (11).

security breach that a report about $\sqcup_p$ in the NYT made $\sqcup$ public  = (1e)
(a report in the NYT)$_{np}$ made (the security breach)$_{np}$ public$_{ap}$

$$\begin{array}{rcl} \text{a, the} & :: & np/n \\ \text{security breach, report, NYT} & :: & n \\ \text{about, in} & :: & (n\backslash n)/np \\ \text{made} & :: & ((np\backslash s)/ap)/np \\ \text{public} & :: & ap \end{array} \qquad (11)$$

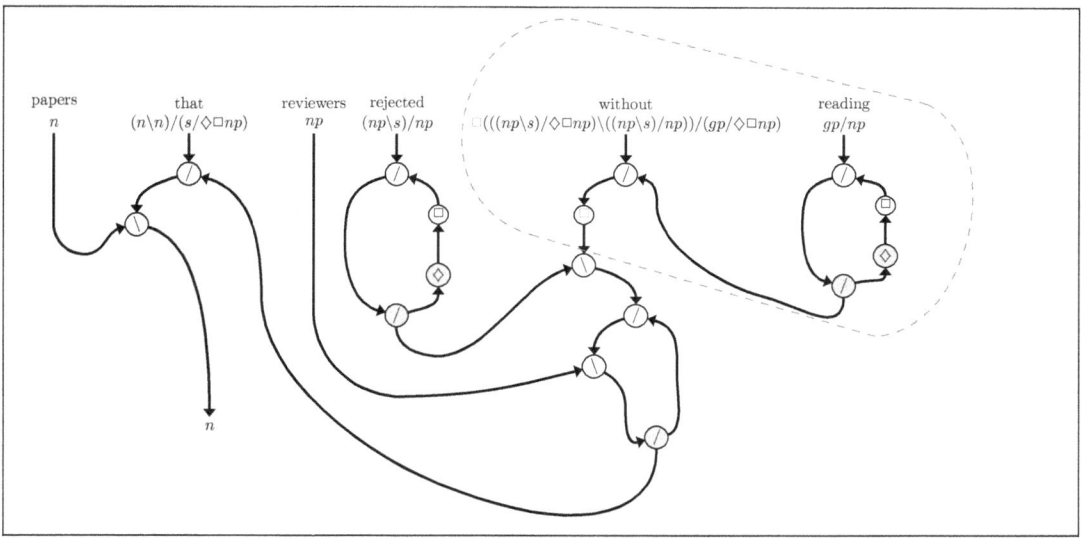

Figure 4: Information flow for the double parasitic gap.

In (1e) the relative clause body does not contain a coordination-like element that would be a suitable candidate to lexically encapsulate the ostensible copying. But we can turn to the relative pronoun itself, and use the mechanisms we relied on for parasitic gaps in adjuncts to move from the relative pronoun's basic type assignment for single-gap dependencies to a derived assignment for the double-gap dependency of (1e).

$$\begin{array}{rcl} \text{that}^{a,c} & :: & (n\backslash n)/(s/\Diamond\Box np) \\ \text{that}^{e} & :: & (n\backslash n)/((np/\Diamond\Box np) \otimes ((np\backslash s)/\Diamond\Box np)) \end{array} \qquad (12)$$

Again, we see that these types are derivable from the initial type for *that* by a combination of an expansion and a distribution step:

$$(n\backslash n)/(s/\Diamond\Box np)$$
$$\downarrow \text{expand}$$
$$(n\backslash n)/((np \otimes np\backslash s)/\Diamond\Box np)$$
$$\downarrow \text{distribute}$$
$$(n\backslash n)/((np/\Diamond\Box np) \otimes ((np\backslash s)/\Diamond\Box np))$$

The expansion step replaces $s$ in antitone position by $np \otimes np\backslash s$, which is justified by leftward Application $ev^\backslash : np \otimes np\backslash s \longrightarrow s$ and Monotonicity. Here, with $A = np \otimes np\backslash s$, $B = s$, $C = \Diamond\Box np$ and $D = n\backslash n$, we have

$$\dfrac{\dfrac{\dfrac{\quad}{A \longrightarrow B}\,\text{Appl}}{A/C \longrightarrow B/C}\,\text{Mon}^\uparrow}{D/(B/C) \longrightarrow D/(A/C)}\,\text{Mon}^\downarrow$$

Likewise, the distribution step relies on Mon$^\downarrow$ to replace $(A \otimes B)/C$ by $A/C \otimes B/C$ in antitone position. Here, with $A = np$, $B = np\backslash s$, $C = \Diamond\Box np$, $D = n\backslash n$, we have

$$\dfrac{\dfrac{\dfrac{\dfrac{\vdots}{(A/C \otimes C) \otimes (B/C \otimes C) \longrightarrow A \otimes B}}{(A/C \otimes B/C) \otimes C \longrightarrow A \otimes B}\,\text{Distr}}{A/C \otimes B/C \longrightarrow (A \otimes B)/C}\,\text{Res}}{D/((A \otimes B)/C) \longrightarrow D/(A/C \otimes B/C)}\,\text{Mon}^\downarrow$$

Figure 6 has the derivation for example (1e).

Turning to (1f), repeated below with its underlying lexical type-assignments, we find the primary and secondary gaps in the infinitival complement *to_inf* and direct object of the verb *persuade*.

candidate whom Alice persuaded every friend of ⊔ to vote for ⊔    ∼ (1f)
Alice$_{np}$ persuaded (a friend)$_{np}$ to vote for Bob$_{np}$

$$\begin{aligned}
\text{persuaded} &\;::\; ((np\backslash s)/to\_inf)/np \\
\text{to vote} &\;::\; to\_inf/pp \\
\text{for} &\;::\; pp/np \\
\text{whom}^f &\;::\; (n\backslash n)/(((s/\Diamond\Box to\_inf)/\Diamond\Box np) \otimes (to\_inf/\Diamond\Box np))
\end{aligned} \qquad (13)$$

To obtain the required derived type for *whom*, we follow the same expansion/distribution routine as for (1f). Expansion in this case replaces $s$ by the product $(s/to\_inf) \otimes to\_inf$; the gap type $\Diamond\Box np$ is then distributed over the two factors of that product. To obtain the desired whom$^f$, there is an extra modal marking on the first occurrence of *to_inf*, in order to license rebracketing with respect to the subject. Recall that the base logic **NL** is non-associative by default.

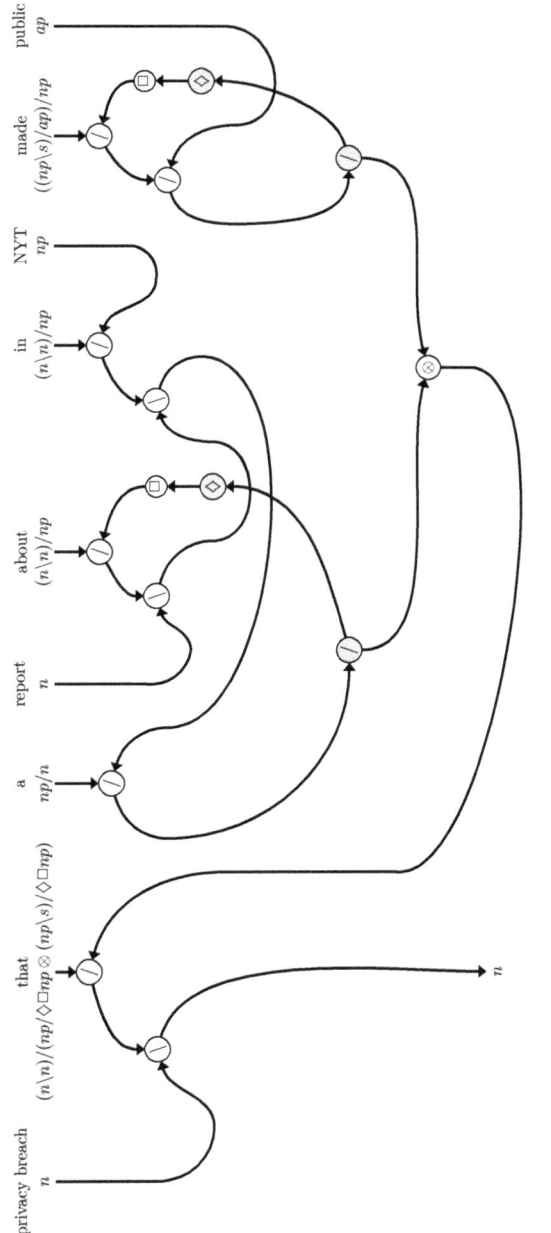

Figure 5: Co-argument parasitic gapping (1e).

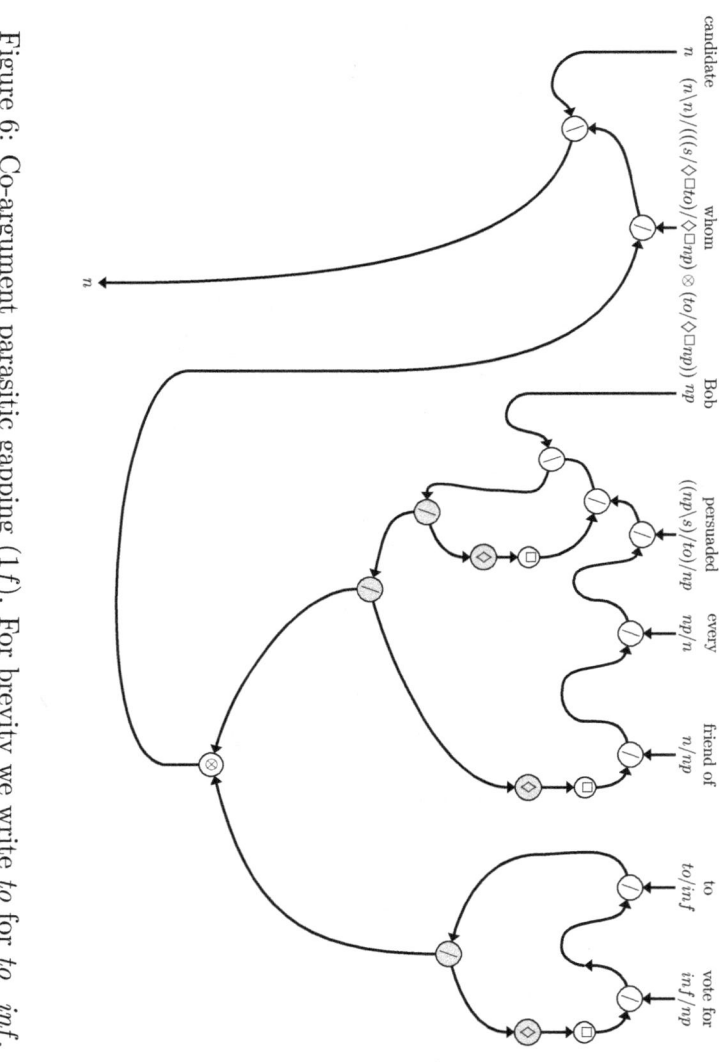

Figure 6: Co-argument parasitic gapping (1f). For brevity we write $to$ for $to\_inf$.

# A Frobenius Algebraic Analysis for Parasitic Gaps

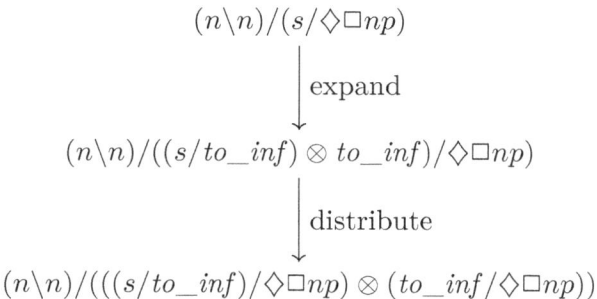

## 3 Frobenius Semantics

The proposed vector-based semantics has two ingredients: first, the *derivational* semantics specifies a compositional mapping that interprets types and proofs of the **NL**$_\diamond$ syntax as morphisms of a Compact Closed Category, concretely the category of **FVect** and linear maps. Second, the *lexical* semantics specifies the word-internal interpretation of individual lexical items; here, we make use of the Frobenius Algebras over **FVect** to model the copying of semantic content associated with the interpretation of relative pronouns such as *that* and *whom*, and modifier heads such as *without*.

### 3.1 Diagrams for Compact Closed Categories and Frobenius Algebras

Recall that a Compact Closed Category is a symmetric monoidal category $(\mathcal{C}, \otimes, I)$ with duals $A^*$ for every object $A$, and *contraction* and *expansion* maps for every object. In the case of vector spaces over fixed bases (our concrete semantics) we don't distinguish between objects and their duals, hence the contraction and expansion maps have signature $\epsilon : V \otimes V \to I$ and $\eta : I \to V \otimes V$, respectively.

For compact closed categories, there is a complete diagrammatic language available, that uses *cups* and *caps* to represent contraction and expansion, see [18]. These are drawn as connecting two objects either as a cup in the case of $\epsilon$ or as a cap in the case of $\eta$. The standard contraction and expansion maps of a CCC form the basis for interpreting derivations of **NL**$_\diamond$.

Crucial to our polymorphic approach is the inclusion of Frobenius Algebras in the lexicon. A Frobenius algebra in a symmetric monoidal category $(\mathcal{C}, \otimes, I)$ is a tuple $(X, \Delta, \iota, \mu, \zeta)$ where, for $X$ an object of $\mathcal{C}$, the first triple below is an internal comonoid and the second one is an internal monoid.

$$(X, \Delta, \iota) \qquad (X, \mu, \zeta)$$

This means that we have a coassociative map $\Delta$ and and its counit $\iota$:

$$\Delta \colon X \to X \otimes X \qquad \iota \colon X \to I$$

and an associative map $\mu$ and its unit $\zeta$:

$$\mu \colon X \otimes X \to X \qquad \zeta \colon I \to X$$

as morphisms of our category $\mathcal{C}$. The $\Delta$ and $\mu$ morphisms satisfy the *Frobenius condition* given below

$$(\mu \otimes 1_X) \circ (1_X \otimes \Delta) = \Delta \circ \mu = (1_X \otimes \mu) \circ (\Delta \otimes 1_X)$$

Informally, the comultiplication $\Delta$ decomposes the information contained in one object into two objects; the multiplication $\mu$ combines the information of two objects into one. In diagrammatic terms, to visualise the Frobenius operations one adds a white triangle to the diagrammatic language for CCCs that represents the (un)merging of information through the four different Frobenius maps. The resulting graphical language is summarised in Figure 7.

## 3.2 Derivational Semantics

For the derivational semantics, we need to define a homomorphism $\lceil \cdot \rceil$ that sends syntactic types and derivations to the corresponding components of the Compact Closed Category of **FVect** and linear maps. This homomorphism has been worked out by Moortgat and Wijnholds [9]. We present the key ingredients below and refer the reader to that paper for full details.

**Types** The target signature has atomic semantic spaces $N$ and $S$, an involutive $(\cdot)^*$ for dual spaces and a symmetric monoidal product $\otimes$. We set

$$\begin{aligned}
\lceil s \rceil &= S, \\
\lceil np \rceil = \lceil n \rceil &= N, \\
\lceil to\_inf \rceil = \lceil ap \rceil = \lceil gp \rceil &= N^* \otimes S, \\
\lceil \Diamond A \rceil = \lceil \Box A \rceil &= \lceil A \rceil, \\
\lceil A/B \rceil &= \lceil A \rceil \otimes \lceil B \rceil^*, \\
\lceil A \backslash B \rceil &= \lceil A \rceil^* \otimes \lceil B \rceil
\end{aligned}$$

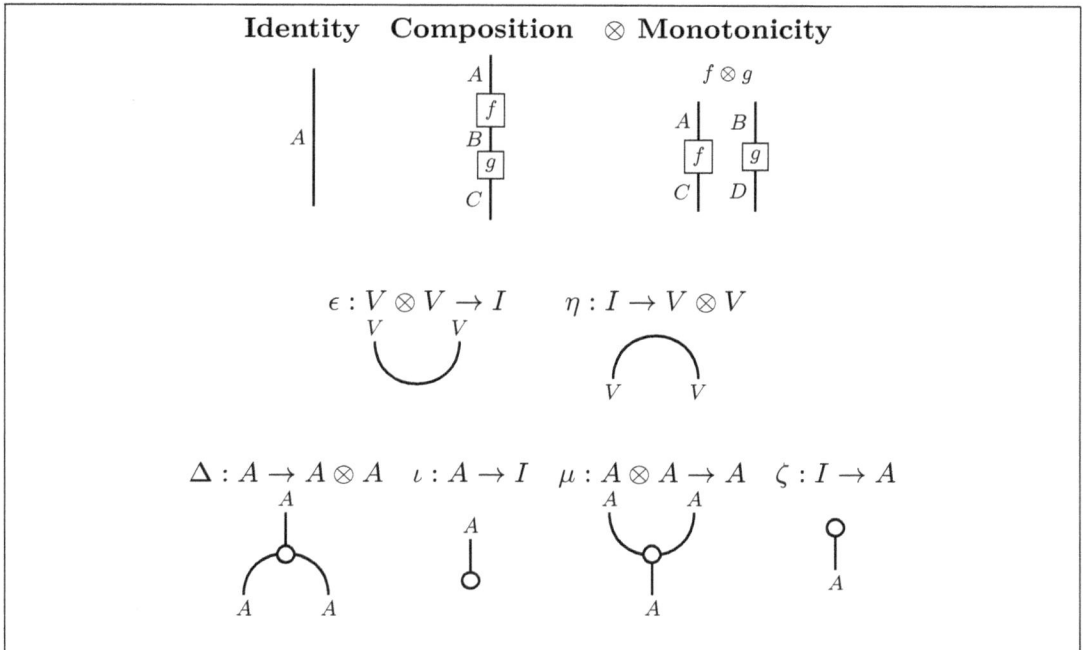

Figure 7: Diagrams of a Compact Closed Category with Frobenius Algebras.

Notice that *to_inf*, *ap* and *gp* are mapped to $N^* \otimes S$. Their understood subject is provided by the context: the main clause subject, in the case of *Bob fell asleep while watching TV*, the direct object in the case of *make the report public* and *persuade A to vote for B*.

**Derivations** The instances of the Evaluation axioms correspond to generalised contraction operations on vector spaces, the instances of the Co-Evaluation axioms dually are mapped to generalised expansion maps. The structural control postulates stipulate a syntactically limited associativity and commutativity; since the control modalities leave no trace on the semantic interpretation, the structural postulates $\alpha_\diamond$ and $\sigma_\diamond$ are interpreted using the standard associativity and symmetry maps of **FVect**.

The derivational semantics is represented graphically in Figure 8, where the diagrams of Figure 2 are interpreted in the complete diagrammatic language of compact closed categories of Figure 7.

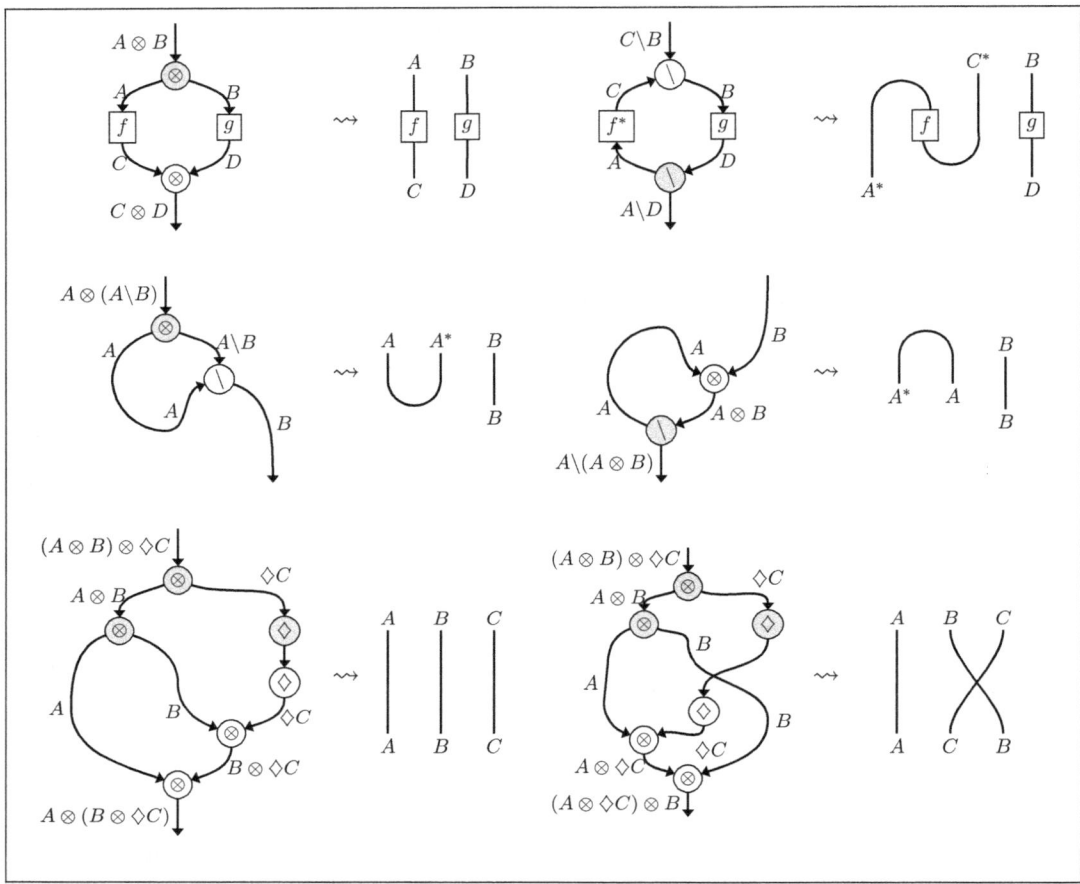

Figure 8: Interpreting derivations of **NL**$_\diamond$ arrows in a compact closed category.

Under the given interpretation, the diagrammatic derivation of Figure 4 for (1d)

| papers | that | Bob | rejected | without | reading | |
|---|---|---|---|---|---|---|
| $n$ | $(n\backslash n)/(s/\diamond\Box np)$ | $np$ | $(np\backslash s)/np$ | $(\Box(X\backslash Y))/Z$ | $gp/np$ | $\longrightarrow n$ |

is sent to the contractions in the interpreting CCC in Figure 9 (red: ⌈that⌉, blue: ⌈without⌉).

## 3.3 Lexical Semantics

For the lexical interpretation of the relative pronouns *that* and *whom* and the conjunctive *without*, we follow previous work [16, 17] and use Frobenius algebras that

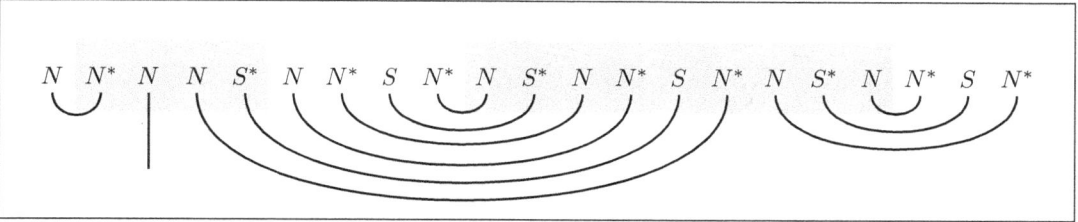

Figure 9: Axiom linking in a CCC for the parasitic gapping example (1d).

characterise vector space bases [1]. First, the basic form of the diagram for *that* is as developed in [16]. The basic diagram for *without* uses a double instance of a Frobenius Algebra to coordinate the gerundive phrase with the intransitive verb phrase consumed to its left. Recall that the interpretation homomorphism sends $np\backslash s$ and $gp$ to the same semantic space, $N^* \otimes S$. In Figure 10 we display graphically these basic types as well as how their *derived* instantiations look. As our type for *whom* is derived similarly to the type of *that*, except that we distribute over the type *to_inf* rather than $np$, we get instead two extra wires rather than a single one.

For the case of parasitic gaps in adjunct positions we use the basic type for *that* and the derived type for *without*. For *that*, its basic Frobenius instantiation has the concrete effect of projecting down the verb phrase into a vector which is consecutively multiplied elementwise with the head noun of the main clause. The diagram for *without* then makes sure to distribute the missing hypothesis of the relative clause over the two gaps in the clause body. Given the identification $\lceil iv \rceil = \lceil gp \rceil$, this is essentially the treatment of coordination of [5].

For the co-argument case, we need make use of the derived type for *that*; its function is now to both specify the need for a clause body missing a hypothetical noun phrase, as well as coordinating this noun phrase through two gaps. Hence, the derived instantiation figures an iterative use of the Frobenius $\mu$ to merge three elements together.

With both the derivational semantics of Figure 9 and the lexical specifications of the constituents of Figure 10 we can put everything together to get the (unnormalised) diagram in Figure 11.

This diagram can be normalised under the equations of the diagrammatic language, leading to the normal form of Figure 12.

The above diagrams are morphisms of a symmetric compact closed category with Frobenius algebras and can be written down in that language as done e.g. in [16, 9]. Here, we provide the closed linear algebraic form of the normal form in Figure 12. For $\overline{\overline{\text{Rejected}}}$ and $\overline{\overline{\text{Not-Reading}}}$ the rank 3 tensors interpreting *rejected* and *(without)*

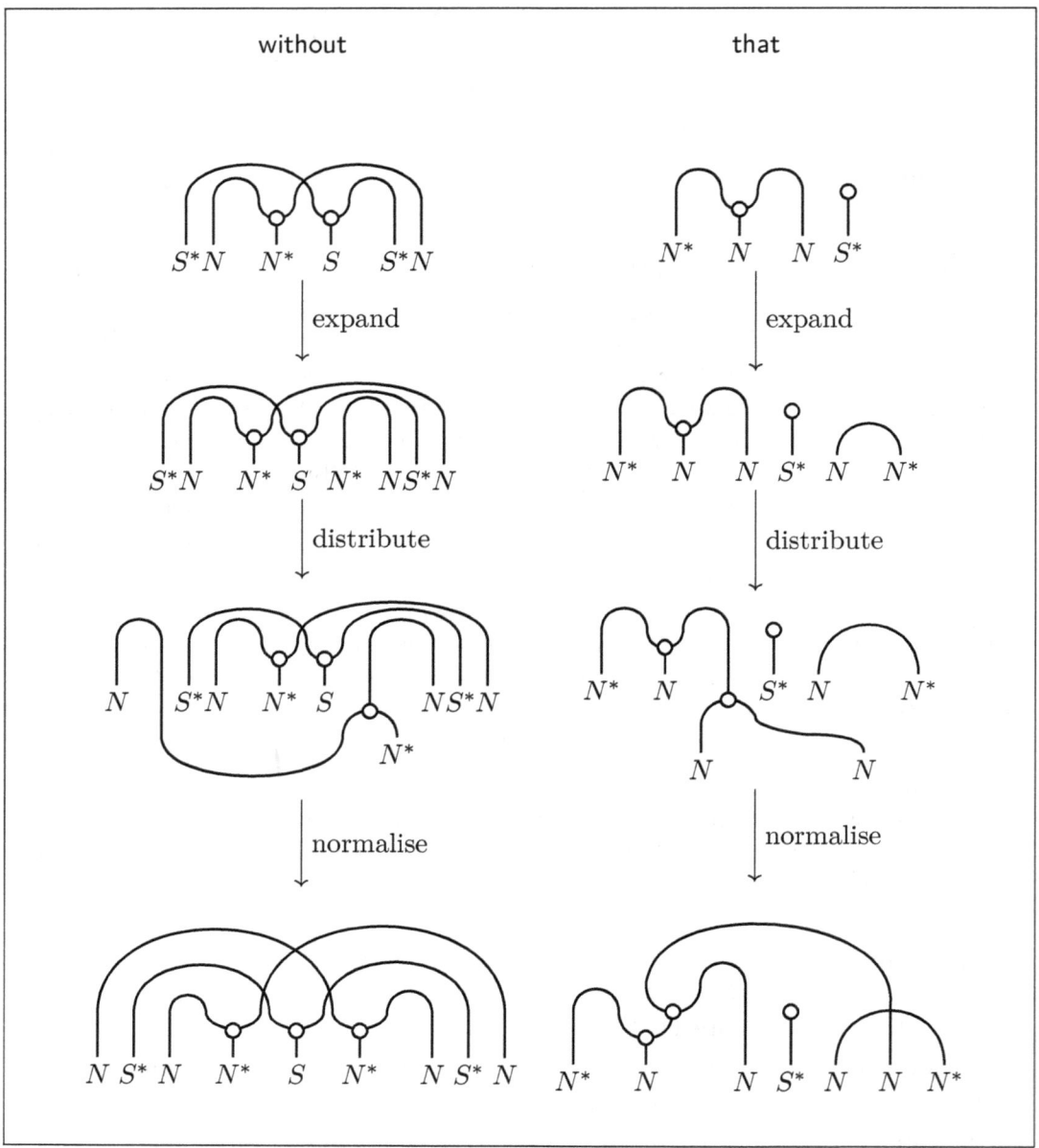

Figure 10: Deriving the lexical semantics for *without* and *that*.

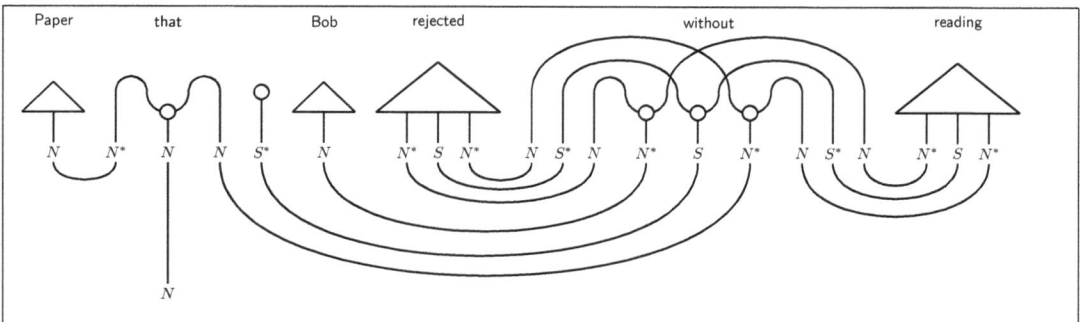

Figure 11: Semantic information flow for the double parasitic gap (initial form).

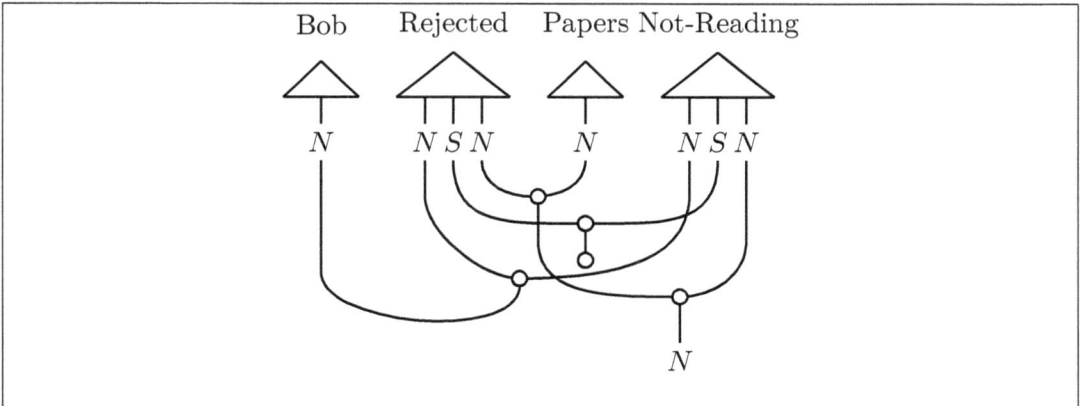

Figure 12: Semantic information flow for the double parasitic gap (normal form).

*reading*, and $\iota$ the unit of the Frobenius coalgebra, this is

$$\overrightarrow{\text{Papers}} \odot (\iota_S \otimes id_N)(\overrightarrow{\text{Bob}}^T \times (\overline{\overrightarrow{\text{Rejected}}} \odot \overline{\overrightarrow{\text{Not-Reading}}}))$$

The closed linear algebraic form says that we take the elementwise multiplication of both cubes, and contract them with the subject *Bob*; then, we collapse the resulting matrix into a vector and compute the elementwise multiplication of this vector with the vector interpreting the head noun *Papers*.

For the co-argument case of parasitic gapping, we insert the derived Frobenius diagrams for *that* and *whom*, to obtain the initial diagrams of Figures 13 (1e) and 14 (1f), which normalise to the diagrams in Figures 15,16. Note that the lexical specification of *made* and *persuade* is a wrapper around the lexical content of the verbs; since *public* and the phrase *to vote for* are interpreted as $N \otimes S$, their understood subject needs to be supplied, which happens through the use of Frobenius operations in the specification of their consuming verbs. This is the direct analogue of assigning a lambda term $\lambda x.\lambda P.\lambda y.PERSUADE\ x\ (P\ x)\ y$ to *persuade*, where the Frobenius expansion corresponds to variable reuse.

## 4 Discussion

The concrete modelling presented above produces an interpretation of relative clauses that is analogous to the formal semantics account: seeing elementwise multiplication as an intersective operation (cf. set intersection), the interpretation of *papers that Bob rejected without reading* identifies those papers that were both rejected and not reviewed, by Bob.

In the formal semantics account, the head noun and the relative clause body are both interpreted as functions from individuals to truth values, i.e. characteristic functions of sets of individuals, which allows them to be combined by set intersection. In our vector-based modelling, however, the head noun and the relative clause body are initially sent to different semantic spaces, viz. $N$ for the head noun versus $N \otimes S$ for the relative clause body. This means we need to appeal to the $\iota$ operation to effectuate the rank reduction from $N \otimes S$ to $N$ that reduces the interpretation of the relative clause body to a vector that can then be conjoined with the meaning of the head noun. The rank reduction performed by the $\iota$ transformation is not a lossless transformation, and it is debatable whether it correctly captures the semantic action we want to associate with the relative pronoun.

As a first step towards a more general model, we abstract away from the specific modelling of the relative pronoun by means of the $\iota$ map.

As shown in Figure 17, our type translation for the relative pronoun effectively interprets it as a map from a verb phrase $(N \otimes S)$ meaning into an adjectival meaning modifying a (common) noun $(N \otimes N)$.

With this generalization, we are not bound anymore to a specific implementation of the relative pronoun meaning, although the proposed account for now gives a workable solution for experimentation.

We suggest here, that a data-driven approach may lend itself for modelling the relative pronoun, as it essentially binds a verb phrase to its adjectival form. For example, a verb phrase can occur in adjectival form, e.g. "papers that were rejected" vs "rejected papers". In such cases, we would expect to get the same meaning representation, which crucially relies on being able to project either an adjective onto a verb phrase or vice versa. Formulating this as a machine learning problem, is work in progress.

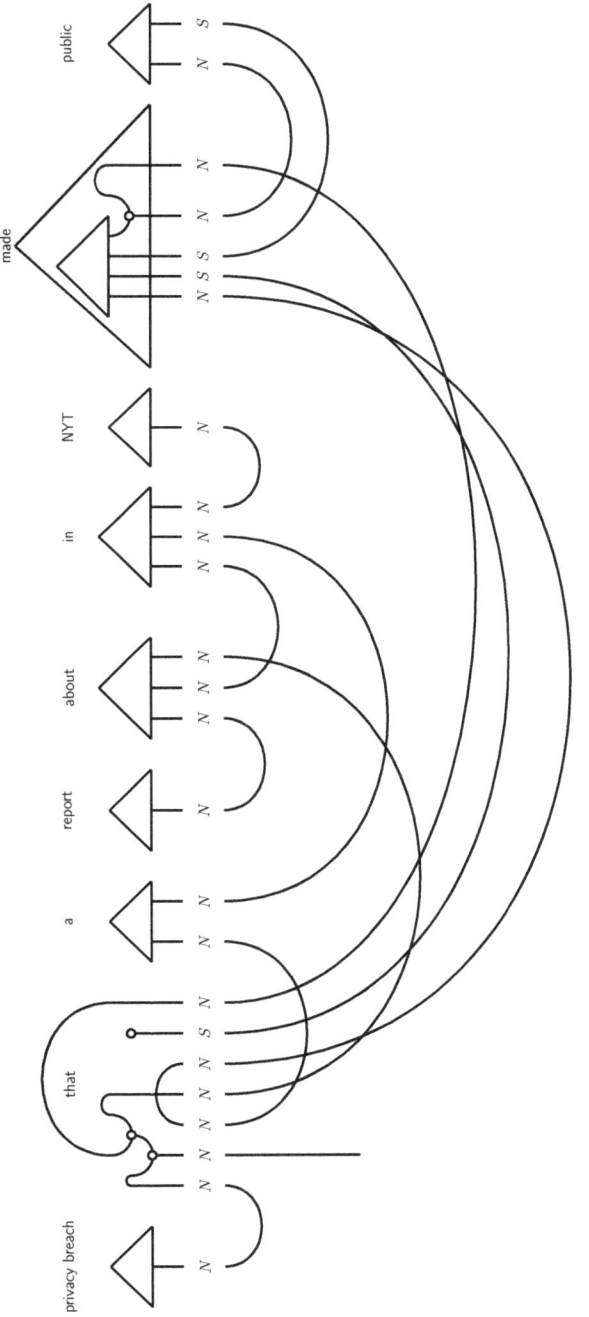

Figure 13: Semantic information flow for the co-argument parasitic gap (1e, initial form).

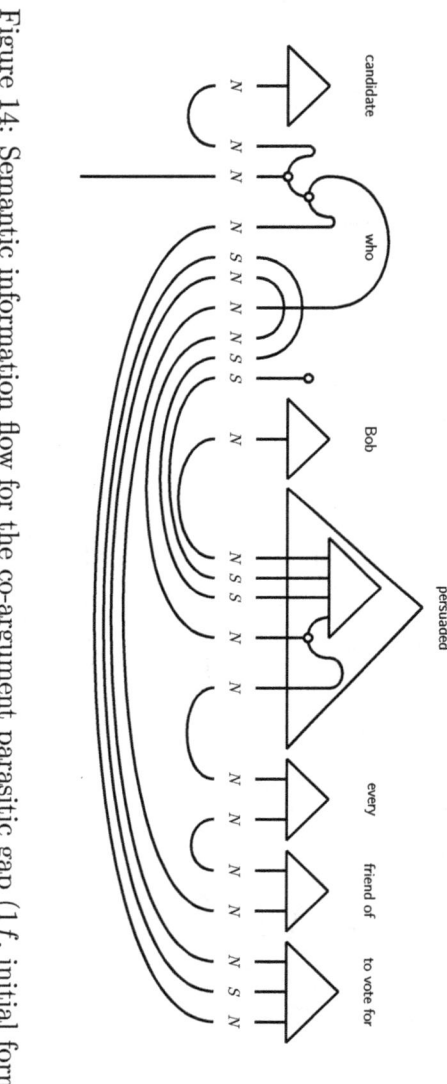

Figure 14: Semantic information flow for the co-argument parasitic gap ($1f$, initial form).

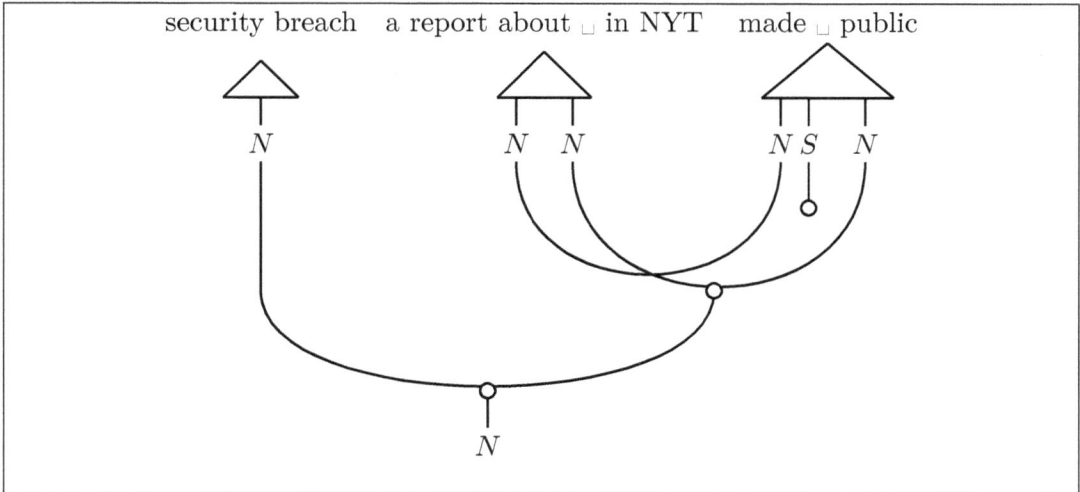

Figure 15: Semantic information flow for the co-argument parasitic gap (1e, normal form).

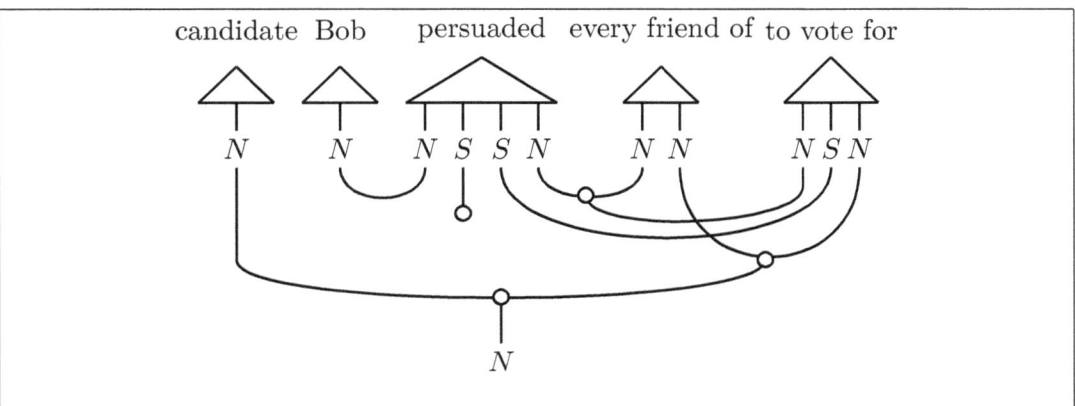

Figure 16: Semantic information flow for the co-argument parasitic gap (1f, normal form).

## 5 Conclusion/Future Work

We presented a typelogical ditributional account of parasitic gapping, one of the many linguistic phenomena in which some semantic elements are not present in the sentence (or more generally discourse) and therefore their corresponding information needs to be provided from some other syntactic element. Rather than relying on some form of copying and/or movement on the syntax side to provide this in-

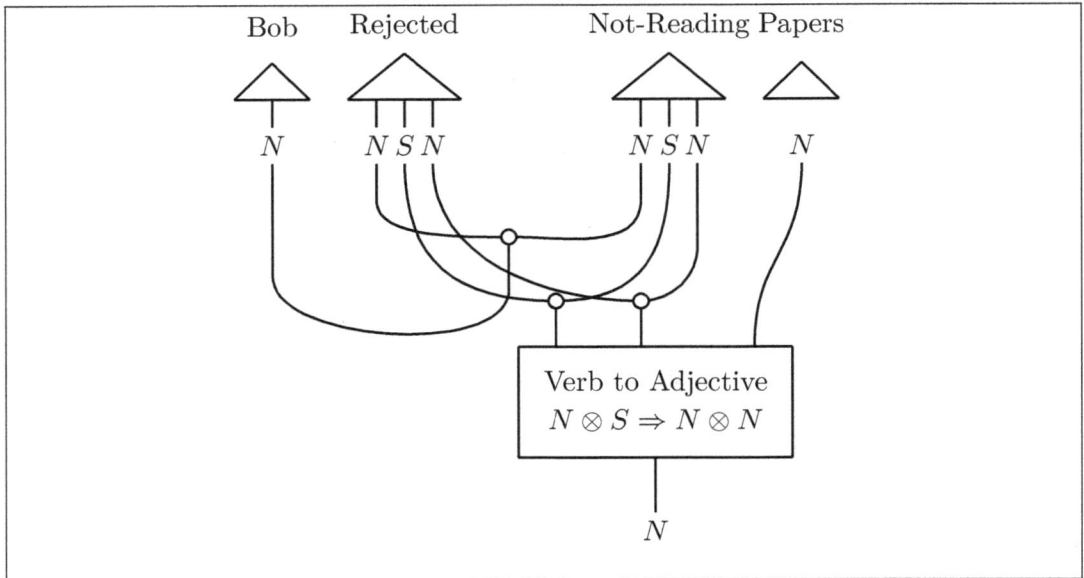

Figure 17: General normal form for a sentence with a parasitic gap; the relative pronoun is now a general map that transforms a verb phrase ($N \otimes S$) into an adjective ($N \otimes N$).

formation (as is the approach for ellipsis with anaphora in [20, 22]), we have solved the problem by using polymorphic typing for function words that play a key role in parasitic gapping (here, *that*, *whom* and *without*).

The polymorphism carries over to the semantics, where we have used Frobenius algebras to interpret them. This enabled us to handle the coordination of multiple gaps, and where the relative pronoun *that* handles the coordination of the head noun with the body of the relative clause and the pronoun *without* coordinates the second gap that exists in the body and which refers to the same head noun. The lexical specifications we use are analogous to a formal semantic modelling, but moreover allow for a more flexible way of representing meaning that may be obtained from data. That resolving gaps is useful in verb disambiguation and sentence similarity tasks has been recently shown [21]. On this point, we discussed a more general normal form in which the behaviour of the relative pronoun is kept abstract. Investigating alternatives to the current modelling with the $\iota$ map, and looking into data-driven modelling of the relative pronoun, constitutes work in progress.

# References

[1] Coecke, B., Pavlovic, D., Vicary, J.: A new description of orthogonal bases. Mathematical Structures in Computer Science 1, 269–272 (2008)

[2] Culicover, P., Postal, P. (eds.): Parasitic Gaps. MIT Press (2001)

[3] Došen, K.: A brief survey of frames for the Lambek calculus. Mathematical Logic Quarterly 38(1), 179–187 (1992)

[4] Kanovich, M.I., Kuznetsov, S., Scedrov, A.: Undecidability of a newly proposed calculus for CatLog3. In: Bernardi, R., Kobele, G., Pogodalla, S. (eds.) Formal Grammar - 24th International Conference, FG 2019, Riga, Latvia, August 11, 2019, Proceedings. Lecture Notes in Computer Science, vol. 11668, pp. 67–83. Springer (2019)

[5] Kartsaklis, D.: Coordination in categorical compositional distributional semantics. In: Proceedings of the 2016 Workshop on Semantic Spaces at the Intersection of NLP, Physics and Cognitive Science, SLPCS@QPL 2016, Glasgow, Scotland, 11th June 2016. pp. 29–38 (2016)

[6] Lambek, J.: The mathematics of sentence structure. American Mathematical Monthly 65, 154–170 (1958)

[7] Lambek, J.: On the calculus of syntactic types. In: Jakobson, R. (ed.) Structure of Language and its Mathematical Aspects, Proceedings of Symposia in Applied Mathematics, vol. XII, pp. 166–178. American Mathematical Society (1961)

[8] Lambek, J.: Type grammar revisited. In: Lecomte, A., Lamarche, F., Perrier, G. (eds.) Logical Aspects of Computational Linguistics, Second International Conference, LACL '97, Nancy, France, September 22-24, 1997, Selected Papers. Lecture Notes in Computer Science, vol. 1582, pp. 1–27. Springer (1997)

[9] Moortgat, M., Wijnholds, G.: Lexical and derivational meaning in vector-based models of relativisation. In: Proceedings of the 21st Amsterdam Colloquium. p. 55. ILLC, University of Amsterdam (2017), https://arxiv.org/abs/1711.11513

[10] Moortgat, M.: Multimodal linguistic inference. Journal of Logic, Language and Information 5(3-4), 349–385 (1996)

[11] Morrill, G.: Type Logical Grammar: Categorial Logic of Signs. Kluwer Academic Publishers, Dordrecht (1994)

[12] Morrill, G.: Parsing/theorem-proving for logical grammar CatLog3. Journal of Logic, Language and Information 28(2), 183–216 (2019)

[13] Morrill, G., Valentín, O.: Computational coverage of TLG: Nonlinearity. In: Proceedings of Third Workshop on Natural Language and Computer Science. vol. 32, pp. 51–63. EasyChair Publications (2015)

[14] Morrill, G., Valentín, O.: On the logic of expansion in natural language. In: Logical Aspects of Computational Linguistics. pp. 228–246. Springer (2016)

[15] Partee, B., Root, M.: Generalized conjunction and type ambiguity. In: Bäuerle, R., Schwarze, C., von Stechow, A. (eds.) Meaning, Use, and Interpretation of Language, pp. 361–383. Walter de Gruyter, Berlin (1983)

[16] Sadrzadeh, M., Clark, S., Coecke, B.: The Frobenius anatomy of word meanings I: subject and object relative pronouns. Journal of Logic and Computation 23(6), 1293–1317 (2013)

[17] Sadrzadeh, M., Clark, S., Coecke, B.: Frobenius anatomy of word meanings 2: possessive relative pronouns. Journal of Logic and Computation 26, 785–815 (2014)

[18] Selinger, P.: A survey of graphical languages for monoidal categories. In: Coecke, B. (ed.) New Structures for Physics, Lecture Notes in Physics, vol. 813, pp. 289–355. Springer, Berlin, Heidelberg (2011)

[19] Steedman, M.: Combinatory grammars and parasitic gaps. Natural Language and Linguistic Theory 5, 403–439 (1987)

[20] Wijnholds, G., Sadrzadeh, M.: Classical copying versus quantum entanglement in natural language: The case of VP-ellipsis. EPTCS 283, 2018, pp. 103-119 pp. 103–119 (2018)

[21] Wijnholds, G., Sadrzadeh, M.: Evaluating composition models for verb phrase elliptical sentence embeddings. In: Proceedings of the 2019 Conference of the North American Chapter of the Association for Computational Linguistics: Human Language Technologies, Volume 1 (Long Papers). Association for Computational Linguistics (2019)

[22] Wijnholds, G., Sadrzadeh, M.: A type-driven vector semantics for ellipsis with anaphora using lambek calculus with limited contraction. Journal of Logic, Language and Information (2019)

[23] Wijnholds, G.J.: Coherent diagrammatic reasoning in compositional distributional semantics. In: International Workshop on Logic, Language, Information, and Computation. pp. 371–386. Springer (2017)

# Vector Spaces as Kripke frames

Giuseppe Greco*
*Utrecht University*

Fei Liang[†]
*School of Philosophy and Social Development, Shandong University*

Michael Moortgat
*Utrecht University*

Alessandra Palmigiano[‡]
*Vrije Universiteit Amsterdam*
*Department of Mathematics and Applied Mathematics, University of Johannesburg*

Apostolos Tzimoulis
*Vrije Universiteit Amsterdam*

## Abstract

In recent years, the compositional distributional approach in computational linguistics has opened the way for an integration of the *lexical* aspects of meaning into Lambek's type-logical grammar program. This approach is based on the observation that a sound semantics for the associative, commutative and unital Lambek calculus can be based on vector spaces by interpreting fusion as the tensor product of vector spaces.

In this paper, we build on this observation and extend it to a 'vector space semantics' for the *general* Lambek calculus, based on *algebras over a field* $\mathbb{K}$ (or

---

*The research of the first and third author is supported by a NWO grant under the scope of the project "A composition calculus for vector-based semantic modelling with a localization for Dutch" (360-89-070).

[†]The research of the second author is supported by the Young Scholars Program of Shandong University (11090089964225).

[‡]The research of the fourth and fifth author is supported by the NWO Vidi grant 016.138.314, the NWO Aspasia grant 015.008.054, and a Delft Technology Fellowship awarded to the fourth author in 2013.

$\mathbb{K}$-algebras), i.e. vector spaces endowed with a bilinear binary product. Such structures are well known in algebraic geometry and algebraic topology, since Lie algebras and Hopf algebras are important instances of $\mathbb{K}$-algebras. Applying results and insights from duality and representation theory for the algebraic semantics of nonclassical logics, we regard $\mathbb{K}$-algebras as 'Kripke frames' the complex algebras of which are complete residuated lattices.

This perspective makes it possible to establish a systematic connection between vector space semantics and the standard Routley-Meyer semantics of (modal) substructural logics.

# 1 Introduction

The extended versions of the Lambek calculus [25, 26] currently used in computational syntax and semantics can be considered as multimodal substructural type logics where residuated families of n-ary fusion operations coexist and interact. Examples are multimodal TLG with modalities for structural control [28], the displacement calculus of [30] which combines concatenation and wrapping operations for the intercalation of split strings, or Hybrid TLG [24], with the non-directional implication of linear logic on top of Lambek's directional implications. For semantic interpretation, these formalisms rely on the Curry-Howard correspondence between derivations in a calculus of semantic types and terms of the lambda calculus that can be seen as recipes for compositional meaning assembly. This view of compositionality addresses *derivational* semantics but remains agnostic as to the choice of semantic spaces for *lexical* items.

Compositional distributional semantics [1, 6, 5, 31] satisfactorily addresses the lexical aspects of meaning while preserving the compositional view on how word meanings are combined into meanings for larger phrases. In [5], the syntax-semantics interface takes the form of a homomorphism from Lambek's syntactic calculus, or its pregroup variant, to the compact closed category of finite dimensional vector spaces and linear maps; [29] have the same target interpretation, but obtain it from the non-associative Lambek calculus extended with a pair of adjoint modal operators allowing for controlled forms of associativity and commutativity in the syntax. The interpretation homomorphism in these approaches typically 'forgets' about syntactic fine-structure, sending Lambek's non-commutative, non-unital syntactic fusion operation to the tensor product of the commutative, associative, unital semantic category, and treating the control modalities as semantically inert.

In this paper we start exploring a more general interpretation of the Lambek fusion in vector spaces. Our starting point is the notion of *algebra over a field* $\mathbb{K}$ (or $\mathbb{K}$-algebra). An algebra over a field $\mathbb{K}$ is a vector space over $\mathbb{K}$ endowed with a bilinear product (cf. Definition 2.2). Algebras over a field can be regarded as Kripke (Routley-Meyer) frames in the following way. The vector space structure of a given $\mathbb{K}$-algebra gives rise to a closure operator on the powerset algebra of its underlying vector space (i.e. the closure operator which associates any set of vectors with the subspace of their linear combinations). The closed sets of this closure operator form a complete non distributive (modular, Arguesian, complemented [23, 22, 12]) lattice which interprets the additive connectives ($\wedge, \vee$) of the Lambek calculus (whenever they are considered). The graph of the bilinear product of the $\mathbb{K}$-algebra, seen as a ternary relation, gives rise to a binary fusion operation on the powerset of the vector space in the standard (Routley-Meyer style) way, and moreover the bilinearity of the $\mathbb{K}$-algebra product guarantees that the closure operator mentioned above is a *nucleus*. This fact makes it possible to endow the set of subspaces of a $\mathbb{K}$-algebra with a residuated lattice structure in the standard way (cf. Section 3). This perspective on $\mathbb{K}$-algebras allows us to introduce a more general vector space semantics for the Lambek calculus (expanded with a unary diamond operator and a unary box operator) which we show to be complete (cf. Section 6), and which lends itself to be further investigated with the tools of unified correspondence [7, 8, 9] and algebraic proof theory [19, 16]. We start developing some instances of *correspondence theory* in this environment, by characterizing the first order conditions on any given (modal) $\mathbb{K}$-algebra corresponding to the validity in its associated (modal) residuated lattice of several identities involving (the diamond and) the Lambek fusion such as commutativity, associativity and unitality. Moreover, using these characterizations, we show that commutativity and associativity fail on the residuated lattice associated with certain well known $\mathbb{K}$-algebras.

## 2 Preliminaries

### 2.1 Algebras over a field

**Definition 2.1** ([27]). *Let $\mathbb{K} = (K, +, \cdot, 0, 1)$ be a field. A vector space over $\mathbb{K}$ is a tuple $\mathbb{V} = (V, +, \cdot, 0)^1$ such that*

*(V1) $+ : V \times V \to V$ is commutative, associative and with unit 0;*

*(V2) $- : V \to V$ is s.t. $u + (-u) = 0$ for any $u \in V$;*

---

[1]We overload notation and use the same symbols for sum, product and the constant 0 both in the field $\mathbb{K}$ and in the vector space $\mathbb{V}$, and rely on the context to disambiguate the reading. Notice that in this axiomatization $-$ is the unary inverse operation and it is considered primitive.

(V2) $\cdot : \mathbb{K} \times V \to V$ *(called the* scalar product*) is an action, i.e.* $\alpha \cdot (\beta \cdot u) = (\alpha \cdot \beta) \cdot u$ *for all* $\alpha, \beta \in \mathbb{K}$ *and every* $u \in V$;

(V3) *the scalar product* $\cdot$ *is* bilinear, *i.e.* $\alpha \cdot (u+v) = (\alpha \cdot u) + (\alpha \cdot v)$ *and* $(\alpha+\beta) \cdot u = (\alpha \cdot u) + (\beta \cdot u)$ *for all* $\alpha, \beta \in \mathbb{K}$ *and all* $u, v \in V$;

(V4) $1 \cdot u = u$ *for every* $u \in V$.

A *subspace* $\mathbb{U}$ of a vector space $\mathbb{V}$ as above is uniquely identified by a subset $U \subseteq V$ which is closed under $+, -, \cdot, 0$.

**Definition 2.2.** *An* algebra over $\mathbb{K}$ *(or* $\mathbb{K}$-algebra*) is a pair* $(\mathbb{V}, \star)$ *where* $\mathbb{V}$ *is a vector space* $\mathbb{V}$ *over* $\mathbb{K}$ *and* $\star : V \times V \to V$ *is bilinear, i.e. left- and right-distributive with respect to the vector sum, and* compatible *with the scalar product:*

(L1$\star$) $u \star (v + w) = (u \star v) + (u \star w)$ *and* $(u + v) \star w = (u \star w) + (v \star w)$ *for all* $u, v, w \in V$;

(L2$\star$) $(\alpha \cdot u) \star (\beta \cdot v) = (\alpha\beta) \cdot (u \star v)$ *for all* $\alpha, \beta \in \mathbb{K}$ *and all* $u, v \in V$.

**Definition 2.3.** *A* $\mathbb{K}$-algebra $(\mathbb{V}, \star)$ *is:*

1. associative *if $\star$ is associative;*

2. commutative *if $\star$ is commutative;*

3. unital *if $\star$ has a unit $1$;*

4. idempotent *if $u = u \star u$ for every $u \in \mathbb{V}$;*

5. monoidal *if $\star$ is associative and unital.*

**Example 2.4.** *Let $\mathbb{R}$ denote the field of real numbers. A well known example of $\mathbb{R}$-algebra is the algebra $(\mathbb{H}, \star_H)$ of quaternions [10], where $\mathbb{H}$ is the 4-dimensional vector space over $\mathbb{R}$, and $\star_H : \mathbb{H} \times \mathbb{H} \to \mathbb{H}$ is the* Hamilton product, *defined on the basis elements $\{\mathbf{e}_1, \mathbf{i}, \mathbf{j}, \mathbf{k}\}$ as indicated in the following table and then extended to $\mathbb{H} \times \mathbb{H}$ by bilinearity as usual. Quaternions are the unique associative 4-dimensional $\mathbb{R}$-algebra fixed by $\mathbf{i}^2 = \mathbf{j}^2 = \mathbf{k}^2 = -\mathbf{e}_1$ and $\mathbf{ijk} = -\mathbf{e}_1$.*

| $\star_H$ | $\mathbf{e}_1$ | $\mathbf{i}$ | $\mathbf{j}$ | $\mathbf{k}$ |
|---|---|---|---|---|
| $\mathbf{e}_1$ | $\mathbf{e}_1$ | $\mathbf{i}$ | $\mathbf{j}$ | $\mathbf{k}$ |
| $\mathbf{i}$ | $\mathbf{i}$ | $-\mathbf{e}_1$ | $\mathbf{k}$ | $-\mathbf{j}$ |
| $\mathbf{j}$ | $\mathbf{j}$ | $-\mathbf{k}$ | $-\mathbf{e}_1$ | $\mathbf{i}$ |
| $\mathbf{k}$ | $\mathbf{k}$ | $\mathbf{j}$ | $-\mathbf{i}$ | $-\mathbf{e}_1$ |

The Hamilton product is monoidal (cf. Definition 2.3)[2] and, notably, not commutative.

**Example 2.5.** *Another well known example is the $\mathbb{R}$-algebra $(\mathbb{O}, \star_o)$ of octonions [10] where $\mathbb{O}$ is the 8-dimensional $\mathbb{R}$-vector space $\mathbb{O}$, and $\star_O : \mathbb{O} \times \mathbb{O} \to \mathbb{O}$ is defined on the basis elements $\mathbf{e}_0, \mathbf{e}_1, \mathbf{e}_2, \mathbf{e}_3, \mathbf{e}_4, \mathbf{e}_5, \mathbf{e}_6, \mathbf{e}_7$ as indicated in the following table.*

| $\star_O$ | $\mathbf{e}_0$ | $\mathbf{e}_1$ | $\mathbf{e}_2$ | $\mathbf{e}_3$ | $\mathbf{e}_4$ | $\mathbf{e}_5$ | $\mathbf{e}_6$ | $\mathbf{e}_7$ |
|---|---|---|---|---|---|---|---|---|
| $\mathbf{e}_0$ | $\mathbf{e}_0$ | $\mathbf{e}_1$ | $\mathbf{e}_2$ | $\mathbf{e}_3$ | $\mathbf{e}_4$ | $\mathbf{e}_5$ | $\mathbf{e}_6$ | $\mathbf{e}_7$ |
| $\mathbf{e}_1$ | $\mathbf{e}_1$ | $-\mathbf{e}_0$ | $\mathbf{e}_3$ | $-\mathbf{e}_2$ | $\mathbf{e}_5$ | $-\mathbf{e}_4$ | $-\mathbf{e}_7$ | $\mathbf{e}_6$ |
| $\mathbf{e}_2$ | $\mathbf{e}_2$ | $-\mathbf{e}_3$ | $-\mathbf{e}_0$ | $\mathbf{e}_1$ | $\mathbf{e}_6$ | $\mathbf{e}_7$ | $-\mathbf{e}_4$ | $-\mathbf{e}_5$ |
| $\mathbf{e}_3$ | $\mathbf{e}_3$ | $\mathbf{e}_2$ | $-\mathbf{e}_1$ | $-\mathbf{e}_0$ | $\mathbf{e}_7$ | $-\mathbf{e}_6$ | $\mathbf{e}_5$ | $-\mathbf{e}_4$ |
| $\mathbf{e}_4$ | $\mathbf{e}_4$ | $-\mathbf{e}_5$ | $-\mathbf{e}_6$ | $-\mathbf{e}_7$ | $-\mathbf{e}_0$ | $\mathbf{e}_1$ | $\mathbf{e}_2$ | $\mathbf{e}_3$ |
| $\mathbf{e}_5$ | $\mathbf{e}_5$ | $\mathbf{e}_4$ | $-\mathbf{e}_7$ | $\mathbf{e}_6$ | $-\mathbf{e}_1$ | $-\mathbf{e}_0$ | $-\mathbf{e}_3$ | $\mathbf{e}_2$ |
| $\mathbf{e}_6$ | $\mathbf{e}_6$ | $\mathbf{e}_7$ | $\mathbf{e}_4$ | $-\mathbf{e}_5$ | $-\mathbf{e}_2$ | $\mathbf{e}_3$ | $-\mathbf{e}_0$ | $-\mathbf{e}_1$ |
| $\mathbf{e}_7$ | $\mathbf{e}_7$ | $-\mathbf{e}_6$ | $\mathbf{e}_5$ | $\mathbf{e}_4$ | $-\mathbf{e}_3$ | $-\mathbf{e}_2$ | $\mathbf{e}_1$ | $-\mathbf{e}_0$ |

*The product of octonions is unital, but neither commutative nor associative.*

**Example 2.6.** *Finally two more examples are the algebras $(\mathbb{M}_n, \star)$, and $(\mathbb{M}_n, \circ_J)$ where $\mathbb{M}_n$ is the vector space of $n \times n$ matrices over $\mathbb{R}$, $\star$ is the usual matrix product and $\circ_J$ is the Jordan product defined as $A \circ_J B = \frac{A \star B + B \star A}{2}$. The usual matrix product is associative but not commutative while the Jordan product is commutative but not associative.*

## 2.2 The modal non associative Lambek calculus

The logic of the modal non associative Lambek calculus **NL**$_\diamond$ can be captured via the *proper* display calculus **D.NL**$_\diamond$ (cf. [32] where this notion is introduced and [19], which expands on the connection between this calculi and the notion of *analytic structural rules*). Notice that the rules of a Gentzen calculus for this logic are derivable in **D.NL**. Moreover, the general theory of display calculi guarantees good properties we want to retain, for instance the fact that any display calculus can be expanded with analytic structural rules still preserving a canonical form of cut-elimination. The language of **D.NL**$_\diamond$ is built from the following structural and operational connectives[3]

---

[2] Given our convention, in this case 1 is an abbreviation for $1\,\mathbf{e}_1 + 0\,\mathbf{i} + 0\,\mathbf{j} + 0\,\mathbf{k}$.
[3] Notice that in [28] the unary modality $\diamond$ is denoted by the symbols $\diamond$ and ∎ is denoted by the symbol $\square^\downarrow$.

| Structural symbols | $\hat{\diamond}$ | $\breve{\blacksquare}$ | $\hat{\otimes}$ | $\breve{/}$ | $\breve{\backslash}$ |
|---|---|---|---|---|---|
| Operational symbols | $\diamond$ | $\blacksquare$ | $\otimes$ | $/$ | $\backslash$ |

The calculus $\mathbf{D.NL}_\diamond$ manipulates formulas and structures defined by the following recursion, where $p \in \mathsf{AtProp}$:

$$\mathsf{Fm} \ni A ::= p \mid \diamond A \mid \blacksquare A \mid A \otimes A \mid A/A \mid A\backslash A$$
$$\mathsf{Str} \ni X ::= A \mid \hat{\diamond} X \mid \breve{\blacksquare} X \mid X \hat{\otimes} X \mid X \breve{/} X \mid X \breve{\backslash} X$$

and consists of the following rules:

**Identity and Cut**

$$\mathrm{Id}\, \frac{}{p \Rightarrow p} \qquad \frac{X \Rightarrow A \quad A \Rightarrow Y}{X \Rightarrow Y}\, \mathrm{Cut}$$

**Display postulates**

$$\otimes \dashv / \,\frac{\dfrac{Y \Rightarrow X \breve{\backslash} Z}{X \hat{\otimes} Y \Rightarrow Z}}{X \Rightarrow Z \breve{/} Y}\, \otimes \dashv \backslash \qquad \diamond \dashv \blacksquare \,\frac{\hat{\diamond} X \Rightarrow Y}{X \Rightarrow \breve{\blacksquare} Y}$$

**Logical rules**

$$\otimes_L \frac{A \hat{\otimes} B \Rightarrow X}{A \otimes B \Rightarrow X} \qquad \frac{X \Rightarrow A \quad Y \Rightarrow B}{X \hat{\otimes} Y \Rightarrow A \otimes B}\, \otimes_R$$

$$\backslash_L \frac{X \Rightarrow A \quad B \Rightarrow Y}{A \backslash B \Rightarrow X \breve{\backslash} Y} \qquad \frac{X \Rightarrow A \breve{\backslash} B}{X \Rightarrow A \backslash B}\, \backslash_R$$

$$/_L \frac{B \Rightarrow Y \quad X \Rightarrow A}{B/A \Rightarrow Y \breve{/} X} \qquad \frac{X \Rightarrow B \breve{/} A}{X \Rightarrow B/A}\, /_R$$

$$\diamond_L \frac{\hat{\diamond} A \Rightarrow X}{\diamond A \Rightarrow X} \qquad \frac{X \Rightarrow A}{\hat{\diamond} X \Rightarrow \diamond A}\, \diamond_R$$

$$\blacksquare_L \frac{A \Rightarrow X}{\blacksquare A \Rightarrow \breve{\blacksquare} X} \qquad \frac{X \Rightarrow \breve{\blacksquare} A}{X \Rightarrow \blacksquare A}\, \blacksquare_R$$

A modal residuated poset is a structure $P = (P, \leq, \otimes, \backslash, /, \diamond, \blacksquare)$ such that $\leq$ is a partial order and for all $x, y, z \in P$

$$x \otimes y \leq z \text{ iff } x \leq z/y \text{ iff } y \leq x\backslash z$$

$$\Diamond x \leq y \text{ iff } x \leq \blacksquare y.$$

The calculus **D.NL$_\Diamond$** is sound and complete with respect to modal residuated posets. Indeed every rule given above is clearly sound on these structures, and the Lindenbaum-Tarski algebra of **D.NL$_\Diamond$** is clearly a modal residuated poset (cf. Proposition 9 and the discussion before Theorem 4 in [16]). Furthermore, **D.NL$_\Diamond$** has the finite model property with respect to modal residuated posets (cf. [16, Theorem 49]).

**Analytic Extensions.** As an example of an extension of **D.NL$_\Diamond$** with analytic structural rules, consider $A\Diamond$ and $\Diamond C$ below.

$$A\Diamond \frac{X \hat{\otimes} (Y \hat{\otimes} \hat{\Diamond} Z) \Rightarrow W}{(X \hat{\otimes} Y) \hat{\otimes} \hat{\Diamond} Z \Rightarrow W} \qquad \Diamond C \frac{(X \hat{\otimes} \hat{\Diamond} Y) \hat{\otimes} Z \Rightarrow W}{(X \hat{\otimes} Z) \hat{\otimes} \hat{\Diamond} Y \Rightarrow W}$$

These rules replace global forms of associativity or commutativity by controlled forms of restructuring ($A\Diamond$) or reordering ($\Diamond C$) that have to be explicitly licensed by the presence of the $\hat{\Diamond}$ operation. Rules of this form have been used to model long range dependencies: constructions where a question word or relative pronoun has to provide the semantic content for an unrealized 'virtual' element later in the phrase. In the relative clause `key that Alice found` ␣ `there`, for instance, the relative pronoun *that* has to make sure that the unrealized direct object of `found` (indicated by ␣) is understood as the key. To make this possible, typological grammars assign a higher-order type to the relative pronoun; the unexpressed object then has the logical status of a *hypothesis* that can be withdrawn once it has been used to provide the transitive verb with its direct object.

We illustrate with the following simple lexicon: `key` : $n$, `that` : $(n\backslash n)/(s/\Diamond\blacksquare np)$, `Alice` : $np$, `found` : $(np\backslash s)/np$, `there` : $(np\backslash s)\backslash(np\backslash s)$. Consider first the judgment `key that Alice found` $\Rightarrow n$ where the gap ␣ occurs at the right periphery of the clause `Alice found` ␣. In the derivations below a dashed inference line abbreviates applications of display postulates or unary logical rules. The derivation relies on controlled associativity $A\Diamond$:

859

$$\cfrac{n \Rightarrow n \qquad n \Rightarrow n}{n \setminus n \Rightarrow n \check{\setminus} n} \qquad A\diamond \cfrac{\cfrac{\cfrac{\cfrac{np \Rightarrow np \qquad s \Rightarrow s}{np \setminus s \Rightarrow np \check{\setminus} s} \qquad \cfrac{np \Rightarrow np}{\hat{\diamond}\blacksquare np \Rightarrow np}}{(np \setminus s)/np \Rightarrow (np \check{\setminus} s) \check{/} \hat{\diamond}\blacksquare np}}{\cfrac{np \,\hat{\otimes}\, ((np \setminus s)/np \,\hat{\otimes}\, \hat{\diamond}\blacksquare np) \Rightarrow s}{(np \,\hat{\otimes}\, (np \setminus s)/np) \,\hat{\otimes}\, \hat{\diamond}\blacksquare np \Rightarrow s}}}{np \,\hat{\otimes}\, (np \setminus s)/np \Rightarrow s / \diamond\blacksquare np}$$

$$\cfrac{(n \setminus n)/(s/\diamond\blacksquare np) \Rightarrow (n \check{\setminus} n) \check{/} (np \,\hat{\otimes}\, (np \setminus s)/np)}{\underbrace{n}_{\text{key}} \,\hat{\otimes}\, \underbrace{((n \setminus n)/(s/\diamond\blacksquare np)}_{\text{that}} \,\hat{\otimes}\, (\underbrace{np}_{\text{Alice}} \,\hat{\otimes}\, \underbrace{(np \setminus s)/np}_{\text{found}})) \Rightarrow n}$$

This example would be derivable also in Lambek's [25] Syntactic Calculus, where associativity is globally available. But consider what happens when an adverb is added at the end. We then have to prove the judgment **key that Alice found there** $\Rightarrow n$ where the gap $\hat{\diamond}\blacksquare np$ occurs in a non-peripheral position. The Syntactic Calculus lacks the expressivity to derive such examples. With the help of controlled commutativity $\diamond C$ (and $A\diamond$) the derivation goes through:

$$\text{[derivation tree with } \diamond C \text{ and } A\diamond \text{ rules, ending in]}$$

$$\underbrace{n}_{\text{key}} \,\hat{\otimes}\, (\underbrace{(n \setminus n)/(s/\diamond\blacksquare np)}_{\text{that}} \,\hat{\otimes}\, (\underbrace{np}_{\text{Alice}} \,\hat{\otimes}\, (\underbrace{((np \setminus s)/np)}_{\text{found}} \,\hat{\otimes}\, \underbrace{(np \setminus s) \setminus (np \setminus s)}_{\text{there}}))) \Rightarrow n$$

The original modal Lambek calculus is single-type. However, it is possible to generalize this framework to proper *multi-type* display calculi, which retain the fundamental properties while allowing further flexibility. Languages with different sorts (also called types in this context) are perfectly admissible and so-called heterogeneous connectives are often considered (e.g. [14, 13, 18, 20, 21, 17, 4]). In particular, we may admit heterogeneous unary modalities where the source and the target of $\diamond$ and $\blacksquare$ do not coincide.

## 3 A Kripke-style analysis of algebras over a field

For any $\mathbb{K}$-algebra $(\mathbb{V}, \star)$, the set $\mathcal{S}(\mathbb{V})$ of subspaces of $\mathbb{V}$ is closed under arbitrary intersections, and hence it is a complete sub $\bigcap$-semilattice of $\mathcal{P}(\mathbb{V})$. Therefore, by basic order-theoretic facts (cf. [11]), $\mathcal{S}(\mathbb{V})$ gives rise to a closure operator $[-]: \mathcal{P}(\mathbb{V}) \to \mathcal{P}(\mathbb{V})$ s.t. $[X] := \bigcap \{\mathbb{U} \in \mathcal{S}(\mathbb{V}) \mid X \subseteq \mathbb{U}\}$ for any $X \in \mathcal{P}(\mathbb{V})$. The elements of $[X]$ can be characterized as *linear combinations* of elements in $X$, i.e. for any $v \in \mathbb{V}$,

$$v \in [X] \quad \text{iff} \quad v = \Sigma_i \alpha_i \cdot x_i.$$

If $(\mathbb{V}, \star)$ is a $\mathbb{K}$-algebra, let $\otimes : \mathcal{P}(\mathbb{V}) \times \mathcal{P}(\mathbb{V}) \to \mathcal{P}(\mathbb{V})$ be defined as follows:

$$X \otimes Y := \{x \star y \mid x \in X \text{ and } y \in Y\} = \{z \mid \exists x \exists y (z = x \star y \text{ and } x \in X \text{ and } y \in Y)\}.$$

**Lemma 3.1.** *If $(\mathbb{V}, \star)$ is a $\mathbb{K}$-algebra, $[-]: \mathcal{P}(\mathbb{V}) \to \mathcal{P}(\mathbb{V})$ is a nucleus on $(\mathcal{P}(\mathbb{V}), \otimes)$, i.e. for all $X, Y \in \mathcal{P}(\mathbb{V})$,*

$$[X] \otimes [Y] \subseteq [X \otimes Y].$$

*Proof.* By definition, $[X] \otimes [Y] = \{u \star v \mid u \in [X] \text{ and } v \in [Y]\}$. Let $u \in [X]$ and $v \in [Y]$, and let us show that $u \star v \in [x \star y \mid x \in X \text{ and } y \in Y]$. Since $u = \Sigma_j \beta_j x_j$ for $x_j \in X$, we can rewrite $u \star v$ as follows: $u \star v = (\Sigma_j \beta_j x_j) \star v = \Sigma_j ((\beta_j x_j) \star v) = \Sigma_j \beta_j (x_j \star v)$; likewise, since $v = \Sigma_k \gamma_k y_k$ for $y_k \in Y$, we can rewrite each $x_j \star v$ as $x_j \star v = x_j \star (\Sigma_k \gamma_k y_k) = \Sigma_k (x_j \star (\gamma_k y_k)) = \Sigma_k \gamma_k (x_j \star y_k)$. Therefore:

$$u \star v = \Sigma_j \beta_j (x_j \star v) = \Sigma_j \beta_j (\Sigma_k \gamma_k (x_j \star y_k)) = \Sigma_j \Sigma_k (\beta_j \gamma_k)(x_j \star y_k),$$

which is a linear combination of elements of $X \otimes Y$, as required. $\square$

Hence, by the general representation theory of residuated lattices [15, Lemma 3.33], Lemma 3.1 implies that the following construction is well defined:[4]

**Definition 3.2.** *If $(\mathbb{V}, \star)$ is a $\mathbb{K}$-algebra, let $\mathbb{V}^+ := (\mathcal{S}(\mathbb{V}), \top, \bot, \wedge, \vee, \otimes, \backslash, /)$ be the complete residuated lattice generated by $(\mathbb{V}, \star)$, i.e. for all $\mathbb{U}, \mathbb{W}, \mathbb{Z} \in \mathcal{S}(\mathbb{V})$,*

$$\mathbb{U} \otimes \mathbb{W} \subseteq \mathbb{Z} \quad \text{iff} \quad \mathbb{U} \subseteq \mathbb{Z}/\mathbb{W} \quad \text{iff} \quad \mathbb{W} \subseteq \mathbb{U}\backslash\mathbb{Z}, \tag{1}$$

*where*

1. $\top := \mathbb{V}$

---

[4] Notice that in defining the operations, we prefer to use the standard universal and existential modal logic clauses associated to left and right residuals, respectively.

2. $\perp := \{0\}$

3. $\mathbb{U} \vee \mathbb{W} := [z \mid \exists u \exists w(z = u + w \text{ and } u \in U \text{ and } w \in W)]$

4. $\mathbb{U} \wedge \mathbb{Z} := \mathbb{U} \cap \mathbb{Z}$

5. $\mathbb{U} \otimes \mathbb{W} := [z \mid \exists u \exists w(z = u \star w \text{ and } u \in U \text{ and } w \in W)]$;

6. $\mathbb{Z}/\mathbb{W} := [u \mid \forall z \forall w((z = u \star w \text{ and } w \in W) \Rightarrow z \in Z)]$;

7. $\mathbb{U}\backslash\mathbb{Z} := [w \mid \forall u \forall z((z = u \star w \text{ and } u \in U) \Rightarrow z \in Z)]$.

**Lemma 3.3.** $[\mathbb{U} \cup \mathbb{W}] = \mathbb{U} \vee \mathbb{W}$.

*Proof.* To show $[\mathbb{U} \cup \mathbb{W}] \subseteq \mathbb{U} \vee \mathbb{W}$, it is enough to show that $\mathbb{U} \cup \mathbb{W} \subseteq \{z \mid \exists u \exists w(z = u + w \text{ and } u \in U \text{ and } w \in W)\}$. Let $x \in \mathbb{U} \cup \mathbb{W}$, which implies $x \in U$ or $x \in W$. Without loss of generality, assume that $x \in U$, the definition of subspace implies that $0 \in \mathbb{W}$. Hence $x \in \{z \mid \exists u \exists w(z = u + w \text{ and } u \in U \text{ and } w \in W)\}$ by the fact that $x = x + 0$. Conversely, to show $\mathbb{U} \vee \mathbb{W} \subseteq [\mathbb{U} \cup \mathbb{W}]$, let $z \in \mathbb{U} \vee \mathbb{W}$, we need to show that $z \in [\mathbb{U} \cup \mathbb{W}]$. Since $z = \Sigma_i \alpha_i (u_i + w_i)$ for all $u_i \in U$ and for all $w_i \in W$, $z = \Sigma_i \alpha_i u_i + \Sigma_i \alpha_i w_i$ for all $u_i \in U$ and for all $w_i \in W$. Moreover, since for all $u_i \in U$ and for all $w_i \in W$, $\Sigma_i \alpha_i u_i \in \mathbb{U} \subseteq \mathbb{U} \cup \mathbb{W}$ and $\Sigma_i \alpha_i w_i \in \mathbb{W} \subseteq \mathbb{U} \cup \mathbb{W}$, $\Sigma_i \alpha_i u_i + \Sigma_i \alpha_i w_i \in [\mathbb{U} \cup \mathbb{W}]$ by the definition of $[-]$. Therefore, $z \in [\mathbb{U} \cup \mathbb{W}]$, as required. □

# 4 Sahlqvist correspondence for algebras over a field

**Definition 4.1.** *If* $(\mathbb{V}, \star)$ *is a* $\mathbb{K}$*-algebra,* $\mathbb{V}^+ = (\mathcal{S}(\mathbb{V}), \leq, \otimes, \backslash, /)$ *is:*

1. associative *if $\otimes$ is associative;*

2. commutative *if $\otimes$ is commutative;*

3. unital *if there exists a 1-dimensional subspace $\mathbb{1}$ such that $\mathbb{U} \otimes \mathbb{1} = \mathbb{U} = \mathbb{1} \otimes \mathbb{U}$ for all $\mathbb{U}$;*

4. contractive *if $\mathbb{U} \subseteq \mathbb{U} \otimes \mathbb{U}$ for all $\mathbb{U}$;*

5. expansive *if $\mathbb{U} \otimes \mathbb{U} \subseteq \mathbb{U}$ for all $\mathbb{U}$;*

6. monoidal *if $\otimes$ is associative and unital.*

The following are to be regarded as first-order conditions on $\mathbb{K}$-algebras, seen as 'Kripke frames'.

**Definition 4.2.** *A $\mathbb{K}$-algebra $(\mathbb{V}, \star)$ is:*

1. quasi-commutative *if $\forall u, v \in \mathbb{V} \, \exists \alpha \in \mathbb{K}$ s.t. $u \star v = \alpha(v \star u)$;*

2. quasi-associative *if $\forall u, v, w \in \mathbb{V} \, \exists \alpha \in \mathbb{K}$ s.t. $(u \star v) \star w = \alpha(u \star (v \star w))$ and $\exists \beta \in \mathbb{K}$ s.t. $u \star (v \star w) = \beta((u \star v) \star w)$;*

3. quasi-unital *if $\exists 1 \in \mathbb{V}$ s.t. $\forall u \in \mathbb{V} \, \exists \alpha, \beta, \gamma, \delta \in \mathbb{K}$ s.t. $u = \alpha(u \star 1)$ and $u \star 1 = \beta u$ and $u = \gamma(1 \star u)$ and $1 \star u = \delta u$;*

4. quasi-contractive *if $\forall u \in \mathbb{V} \, \exists \alpha \in \mathbb{K}$ s.t. $u = \alpha(u \star u)$;*

5. quasi-expansive *if $\forall u, v \in \mathbb{V} \, \exists \alpha, \beta \in \mathbb{K}$ s.t. $u \star v = \alpha u + \beta v$;*

6. quasi-monoidal *if quasi-associative and quasi-unital.*

**Remark 4.3.** *The notion of quasi-commutativity is strictly weaker than the notion of commutativity in case $\mathbb{K}$ has more than 2 elements. Indeed take the 2-dimensional vector space over $\mathbb{K}$ with base $e_1, e_2$, and define the bilinear map such that $e_1 \star e_2 = e_1$, $e_2 \star e_1 = -e_1$ and $e_1 \star e_1 = 0 = e_2 \star e_2$. Then it is routine to verify that this $\mathbb{K}$-algebra is quasi-commutative but not commutative.*

In what follows, we sometimes abuse notation and identify a $\mathbb{K}$-algebra $(\mathbb{V}, \star)$ with its underlying vector space $\mathbb{V}$. Making use of definition 4.2 we can show the following:

**Proposition 4.4.** *For every $\mathbb{K}$-algebra $\mathbb{V}$,*

1. *$\mathbb{V}^+$ is commutative iff $\mathbb{V}$ is quasi-commutative;*

2. *$\mathbb{V}^+$ is associative iff $\mathbb{V}$ is quasi-associative;*

3. *$\mathbb{V}^+$ is unital iff $\mathbb{V}$ is quasi-unital;*

4. *$\mathbb{V}^+$ is contractive iff $\mathbb{V}$ is quasi-contractive;*

5. *$\mathbb{V}^+$ is expansive iff $\mathbb{V}$ is quasi-expansive;*

6. *$\mathbb{V}^+$ monoidal iff $\mathbb{V}$ is quasi-monoidal.*

*Proof.* 1. For the left-to-right direction, assume that $\mathbb{V}^+$ is commutative and let $u, v \in \mathbb{V}$. Then $[u] \otimes [v] = [v] \otimes [v]$. Notice that $[u] \otimes [v] = [u \star v] = \{\alpha(u \star v) \mid \alpha \in \mathbb{K}\}$ and $[v] \otimes [u] = [v \star u] = \{\alpha(v \star u) \mid \alpha \in \mathbb{K}\}$. Hence, $[u] \otimes [v] = [v] \otimes [v]$ implies that $u \star v \in [v \star u]$, i.e. $u \star v = \alpha(v \star u)$ for some $\alpha \in \mathbb{K}$, as required.

Conversely, assume that $\mathbb{V}$ is quasi-commutative, and let $\mathbb{U}, \mathbb{W} \in \mathcal{S}(\mathbb{V})$. To show that $\mathbb{U} \otimes \mathbb{W} \subseteq \mathbb{W} \otimes \mathbb{U}$, it is enough to show that $u \star w \in \mathbb{W} \otimes \mathbb{U}$ for every $u \in \mathbb{U}$ and $w \in \mathbb{W}$. By the assumption that $\mathbb{V}$ is quasi-commutative, there exists some $\alpha \in \mathbb{K}$ such that $u \star w = \alpha(w \star u) \in \mathbb{W} \otimes \mathbb{U}$, as required. The argument for $\mathbb{W} \otimes \mathbb{U} \subseteq \mathbb{U} \otimes \mathbb{W}$ is similar, and omitted.

2. For the left-to-right direction, assume that $\mathbb{V}^+$ is associative and let $u, w, z \in \mathbb{V}$. Then $([u] \otimes [w]) \otimes [z] = [u] \otimes ([w] \otimes [z])$. Notice that $([u] \otimes [w]) \otimes [z] = [u \star w] \otimes [z] = [(u \star w) \star z] = \{\alpha((u \star w) \star z) \mid \alpha \in \mathbb{K}\}$ and $[u] \otimes ([w] \otimes [z]) = [u] \otimes [w \star z] = [u \star (w \star z)] = \{\alpha(u \star (w \star z)) \mid \alpha \in \mathbb{K}\}$. Hence, $([u] \otimes [w]) \otimes [z] = [u] \otimes ([w] \otimes [z])$ implies that $(u \star w) \star z = \alpha(u \star (w \star z))$ for some $\alpha \in \mathbb{K}$ and $u \star (w \star z) = \alpha((u \star w) \star z)$ for some $\alpha \in \mathbb{K}$, as required.

Conversely, assume that $\mathbb{V}$ is quasi-associative, and let $\mathbb{U}, \mathbb{W}, \mathbb{Z} \in \mathcal{S}(\mathbb{V})$. To show that $(\mathbb{U} \otimes \mathbb{W}) \otimes \mathbb{Z} \subseteq \mathbb{U} \otimes (\mathbb{W} \otimes \mathbb{Z})$, it is enough to show that $(u \star w) \star z \in \mathbb{U} \otimes (\mathbb{W} \otimes \mathbb{Z})$ for every $u \in \mathbb{U}$, $w \in \mathbb{W}$ and $z \in \mathbb{Z}$. Since $\mathbb{V}$ is quasi-associative, there exists some $\alpha \in \mathbb{K}$ such that $(u \star w) \star z = \alpha(u \star (w \star z)) \in \mathbb{U} \otimes (\mathbb{W} \otimes \mathbb{Z})$, as required. The argument for $\mathbb{U} \otimes (\mathbb{W} \otimes \mathbb{Z}) \subseteq (\mathbb{U} \otimes \mathbb{W}) \otimes \mathbb{Z}$ is similar, and omitted.

3. For the left-to-right direction, assume that $\mathbb{V}^+$ is unital and let $1 \in \mathbb{V}$ such that $\mathbb{1} = [1]$. Then $[u] = [u] \otimes \mathbb{1} = [u \star 1]$ for any $u \in \mathbb{V}$. Hence, $u = \alpha(u \star 1)$ and $u \star 1 = \beta u$, for some $\alpha, \beta \in \mathbb{K}$, as required. Analogously, from $[u] = \mathbb{1} \otimes [u]$ one shows that $u = \gamma(1 \star u)$ and $1 \star u = \delta u$ for some $\gamma, \delta \in \mathbb{K}$.

Conversely, assume that $\mathbb{V}$ is quasi-unital, and let $\mathbb{U} \in \mathcal{S}(\mathbb{V})$. To show that $\mathbb{U} \otimes \mathbb{1} \subseteq \mathbb{U}$, it is enough to show that $u \star 1 \in \mathbb{U}$ for every $u \in \mathbb{U}$. By assumption, there exists some $\alpha \in \mathbb{K}$ such that $u \star 1 = \alpha u \in \mathbb{U}$, as required. The remaining inclusions are proven with similar arguments which are omitted.

4. For the left-to-right direction, assume that $\mathbb{V}^+$ is contractive and let $u \in \mathbb{V}$. Then $[u] \subseteq [u] \otimes [u] = [u \star u]$. Hence, $u = \alpha(u \star u)$ for some $\alpha \in \mathbb{K}$, as required.

Conversely, assume that $\mathbb{V}$ is quasi-contractive, and let $\mathbb{U} \in \mathcal{S}(\mathbb{V})$. To show that $\mathbb{U} \subseteq \mathbb{U} \otimes \mathbb{U}$, it is enough to show that $u \in \mathbb{U} \otimes \mathbb{U}$ for every $u \in \mathbb{U}$. By assumption, there exists some $\alpha \in \mathbb{K}$ such that $u = \alpha(u \star u) \in \mathbb{U} \otimes \mathbb{U}$, as required.

5. For the left-to-right direction, assume that $\mathbb{V}^+$ is expansive and let $u, v \in \mathbb{V}$. Then, letting $[u, v]$ denote the subspace generated by $u$ and $v$, we have $[u, v] \otimes [u, v] \subseteq [u, v]$, and since $u \star v \in [u, v] \otimes [u, v]$ we conclude $u \star v \in [u, v]$, i.e. $u \star v = \alpha u + \beta v$ for some $\alpha, \beta \in \mathbb{K}$, as required.

Conversely, assume that $\mathbb{V}$ is quasi-expansive, and let $\mathbb{U} \in \mathcal{S}(\mathbb{V})$. To show that $\mathbb{U} \otimes \mathbb{U} \subseteq \mathbb{U}$, it is enough to show that $u \star v \in \mathbb{U}$ for every $u, v \in \mathbb{U}$. By assumption, there exist some $\alpha, \beta \in \mathbb{K}$ such that $u \star v = \alpha u + \beta v \in \mathbb{U}$, as required.

6. Immediately follows from 2. and 3. □

## 4.1 Examples

**Fact 4.5.** *The algebra of quaternions $\mathbb{H}$ is not quasi-commutative.*

*Proof.* Let $u = \mathbf{i} + \mathbf{j}$ and $v = \mathbf{j}$, then $u \star_H v = \mathbf{k} - 1$ and $v \star_H u = -\mathbf{k} - 1$. By contradiction, let us assume that $\star_H$ is quasi-commutative, then there exists a real number $\alpha$ s.t. $\mathbf{k} - 1 = \alpha(-\mathbf{k} - 1) = \alpha(-\mathbf{k}) - \alpha$. It follows that $\alpha = 1$ and $a = -1$ contradicting the assumption that $\star_H$ is quasi-commutative. □

**Corollary 4.6.** $\mathbb{H}^+$ *is not commutative.*

*Proof.* Immediate by Fact 4.5 and Proposition 4.4. □

**Fact 4.7.** *The algebra $\mathbb{O}$ of octonions is not quasi-associative.*

*Proof.* Let $u = v = w = 1\,\mathbf{e}_0 + 2\,\mathbf{e}_1 + 3\,\mathbf{e}_2 + 5\,\mathbf{e}_3 + 7\,\mathbf{e}_4 + 8\,\mathbf{e}_5 + 11\,\mathbf{e}_6 + 12\,\mathbf{e}_7$, then $w \star_O u = u \star_O v = -415\,\mathbf{e}_0 + 4\,\mathbf{e}_1 + 6\,\mathbf{e}_2 + 10\,\mathbf{e}_3 + 14\,\mathbf{e}_4 + 96\,\mathbf{e}_5 + 22\,\mathbf{e}_6 + 24\,\mathbf{e}_7$. In order to show that $w \star_O (u \star_O v) \neq (w \star_O u) \star_O v$ is enough to check the first two coordinates: $w \star_O (u \star_O v) = -1887\,\mathbf{e}_0 - 266\,\mathbf{e}_1 \ldots \neq -1887\,\mathbf{e}_0 - 1386\,\mathbf{e}_1 \ldots = (w \star_O u) \star_O v$. By contradiction, let us assume that $\star_O$ is quasi-associative, then there exists a real number $\alpha$ s.t. $w \star_O (u \star_O v) = \alpha((w \star_O u) \star_O v)$. It follows that $-1887\,\mathbf{e}_0 = \alpha(-1887)$ and $-266 = \alpha(-1386)$. We observe that $-1887\,\mathbf{e}_0 = \alpha(-1887)$ holds only for $\alpha = 1$, but then $-266 = \alpha(-1386)$ does not hold contradicting the assumption that $\star_O$ is quasi-associative. □

**Corollary 4.8.** $\mathbb{O}^+$ *is not associative.*

*Proof.* Immediate by Fact 4.7 and Proposition 4.4. □

# 5 Modal algebras over a field

**Definition 5.1.** *A modal $\mathbb{K}$-algebra is a triple $(\mathbb{V}, \star, R)$ such that $(\mathbb{V}, \star)$ is a $\mathbb{K}$-algebra and $R \subseteq V \times V$ is compatible with the scalar product, and it preserves the zero-vector:*

*(L1R)* $vRu \,\&\, zRw \;\Rightarrow\; \forall \gamma \delta \,\exists \alpha \beta \;\; (\gamma v + \delta z) R(\alpha u + \beta w)$;

*(L2R)* $tR(\alpha u + \beta v) \;\Rightarrow\; \exists \lambda \mu \,\exists zw \;\; zRu \,\&\, wRv \,\&\, \lambda z + \mu w = t$.

*(L3R)* $xR0 \;\Leftrightarrow\; x = 0$.

If $(\mathbb{V}, \star, R)$ is a modal $\mathbb{K}$-algebra, let $\Diamond : \mathcal{P}(\mathbb{V}) \to \mathcal{P}(\mathbb{V})$ be defined as follows:

$$\Diamond X := R^{-1}[X] = \{v \mid \exists u(vRu \text{ and } u \in X)\}.$$

**Lemma 5.2.** *If $(\mathbb{V}, R)$ is a modal $\mathbb{K}$-algebra, $[-] : \mathcal{P}(\mathbb{V}) \to \mathcal{P}(\mathbb{V})$ is a $\Diamond$-nucleus on $(\mathcal{P}(\mathbb{V}), \Diamond)$, i.e. for all $X \in \mathcal{P}(\mathbb{V})$,*

$$\Diamond[X] \subseteq [\Diamond X].$$

*Proof.* By definition, $\Diamond[X] = \bigcup \{R^{-1}[u] \mid u \in [X]\}$. Let $u \in [X]$, assume that $vRu$ and let us show that $v \in [\Diamond X]$. Since $u = \Sigma_j \beta_j x_j$ for $x_j \in X$, by L2R, $\forall j \exists \lambda_j \exists v_j \; v_j R x_j \; \& \; \Sigma_j \lambda_j v_j = v$. So $v \in [\Diamond X]$. If $X = \varnothing$, then $\Diamond[\varnothing] = \Diamond \{0\} = R^{-1}[0]$. By L3R, $R^{-1}[0] = \{0\} \subseteq [\Diamond X]$. □

Hence, by the generalization of the representation theory of residuated lattices [2, 3], Lemma 5.2 implies that the following construction is well defined:

**Definition 5.3.** *If $(\mathbb{V}, \star, R)$ is a modal $\mathbb{K}$-algebra, let $\mathbb{V}^+ := (\mathcal{S}(\mathbb{V}), \leq, \otimes, \backslash, /, \Diamond, \blacksquare)$ be the complete modal residuated lattice generated by $(\mathbb{V}, \star, R)$, i.e. for all $\mathbb{U}, \mathbb{W} \in \mathcal{S}(\mathbb{V})$,*

$$\Diamond \mathbb{U} \subseteq \mathbb{W} \quad \text{iff} \quad \mathbb{U} \subseteq \blacksquare \mathbb{W}, \tag{2}$$

*where*

1. $\Diamond \mathbb{U} := [v \mid \exists u \, (vRu \text{ and } u \in U)]$;
2. $\blacksquare \mathbb{W} := [u \mid \forall v \, (vRu \Rightarrow v \in W)]$.

**Remark 5.4.** *Notice that every linear map $f : \mathbb{V} \to \mathbb{V}$ satisfies the conditions of Definition 5.1, and hence functional modal $\mathbb{K}$-algebras $(\mathbb{V}, \star, f)$ can be defined analogously to definition 5.1 and their associated algebras will be complete modal residuated lattices such that $\Diamond f[-] \dashv f^{-1}[-]$ in $\mathcal{S}(\mathbb{V})$. However, if we make use a linear function $f$ (instead of a relation $R$) to define modal $\mathbb{K}$-algebras, then we are not able to show completeness for the full fragment of $\mathbf{D.NL}_\Diamond$.*

## 5.1 Axiomatic extensions of a modal algebra over $\mathbb{K}$

In order to capture controlled forms of associativity/commutativity, we want to consider axiomatic extensions of the modal algebras introduced in the previous section. Below, we consider right-associativity and left-commutativity.

**Definition 5.5.** *If $(\mathbb{V}, \star, f)$ is a modal $\mathbb{K}$-algebra, $\mathbb{V}^+ := (\mathcal{S}(\mathbb{V}), \leq, \otimes, \backslash, /, \Diamond, \blacksquare)$ is:*

1. right-associative *if* $(\mathbb{U} \otimes \mathbb{W}) \otimes \Diamond \mathbb{V} \subseteq \mathbb{U} \otimes (\mathbb{W} \otimes \Diamond \mathbb{V})$;

2. left-commutative *if* $(\mathbb{U} \otimes \mathbb{V}) \otimes \Diamond \mathbb{W} \subseteq (\mathbb{U} \otimes \Diamond \mathbb{W}) \otimes \mathbb{V}$.

**Definition 5.6.** *A modal* $\mathbb{K}$-*algebra* $(\mathbb{V}, \star, R)$ *is:*

1. quasi right-associative *if for* $u, w, z, v \in \mathbb{V}$ *such that* $vRz$, *there exists* $\alpha, \beta \in \mathbb{K}$ *and* $v'$ *such that* $v'R\beta z$ *and* $(u \star w) \star v = \alpha(u \star (w \star v'))$;

2. quasi left-commutative *if for all* $u, w, z, v \in \mathbb{V}$ *such that* $vRz$ *there exists* $\alpha, \beta \in \mathbb{K}$ *and* $v'$ *such that* $v'R\beta z$ *and* $(u \star w) \star v = \alpha((u \star v') \star w)$.

**Proposition 5.7.** *For every modal* $\mathbb{K}$-*algebra* $(\mathbb{V}, \star, R)$:

1. $\mathbb{V}^+$ *is right associative if and only if* $(\mathbb{V}, \star, R)$ *is quasi right-associative;*

2. $\mathbb{V}^+$ *is left commutative if and only if* $(\mathbb{V}, \star, R)$ *is quasi left-commutative.*

*Proof.* 1. For the left to right direction let $u, w, z, v$ such that $vRz$. By the assumption $([u] \otimes [w]) \otimes R^{-1}[[z]] \subseteq [u] \otimes ([w] \otimes R^{-1}[[z]])$. Since $(u \star w) \star v \in ([u] \otimes [w]) \otimes R^{-1}[[z]]$ it follows that $(u \star w) \star v \in [u] \otimes ([w] \otimes R^{-1}[[z]])$, i.e. there exist $\alpha, \beta \in \mathbb{K}$ and $v' \in \mathbb{V}$ with $v'R\beta z$ such that $(u \star w) \star v = \alpha(u \star (w \star v'))$.

For right to left direction let $q \in (\mathbb{U} \otimes \mathbb{W}) \otimes \Diamond \mathbb{Z}$, i.e. there exists $u \in \mathbb{U}, w \in \mathbb{W}$ and $v \in \Diamond \mathbb{Z}$ such that $q = (u \star w) \star v$. Since $v \in \Diamond \mathbb{Z}$ there exists $z \in \mathbb{Z}$ such that $vRz$. Then by assumption there exist $\alpha, \beta \in \mathbb{K}$ and $v' \in \mathbb{V}$ such that $v'R\beta z$ and $q = \alpha(u \star (w \star v'))$. It holds that $v' \in \Diamond \mathbb{Z}$ since $\beta z \in \mathbb{Z}$, and hence $q \in \mathbb{U} \otimes (\mathbb{W} \otimes \Diamond \mathbb{Z})$.

2. For the left to right direction let $u, w, z, v$ such that $vRz$. By the assumption $([u] \otimes [w]) \otimes R^{-1}[[z]] \subseteq ([u] \otimes R^{-1}[[z]]) \otimes [w]$. Since $(u \star w) \star v \in ([u] \otimes [w]) \otimes R^{-1}[[z]]$ it follows that $(u \star w) \star v \in ([u] \otimes R^{-1}[[z]]) \otimes [w]$, i.e. there exist $\alpha, \beta \in \mathbb{K}$ and $v' \in \mathbb{V}$ with $v'R\beta z$ such that $(u \star w) \star v = \alpha((u \star v') \star w)$.

For right to left direction let $q \in (\mathbb{U} \otimes \mathbb{W}) \otimes \Diamond \mathbb{Z}$, i.e. there exist $u \in \mathbb{U}, w \in \mathbb{W}$ and $v \in \Diamond \mathbb{Z}$ such that $q = (u \star w) \star v$. Since $v \in \Diamond \mathbb{Z}$ there exists $z \in \mathbb{Z}$ such that $vRz$. Then by assumption there exists $\alpha, \beta \in \mathbb{K}$ and $v' \in \mathbb{V}$ such that $v'R\beta z$ and $q = \alpha((u \star v') \star w)$. It holds that $v' \in \Diamond \mathbb{Z}$ since $\beta z \in \mathbb{Z}$, and hence $q \in (\mathbb{U} \otimes \Diamond \mathbb{Z}) \otimes \mathbb{W}$. □

**Remark 5.8.** *Notice that in case $R$ is a linear function, the inequalities above imply equality. Indeed, e.g. in the case of right-associativity, if $zRv$, and $\beta zRv'$ then $v' = \beta v$. Therefore, it immediately follows that $(u \star w) \star v = \alpha(u \star (w \star v))$, and hence $\frac{1}{\alpha}((u \star w) \star v) = u \star (w \star v)$, and hence $\mathbb{U} \otimes (\mathbb{W} \otimes \Diamond \mathbb{Z}) \subseteq (\mathbb{U} \otimes \mathbb{W}) \otimes \Diamond \mathbb{Z}$.*

## 6 Completeness

The aim of this section is to show the completeness of the logic $\mathbf{D.NL_\diamond}$ with respect to modal $\mathbb{K}$-algebras of finite dimension (cf. Theorem 6.1).

Given a modal $\mathbb{K}$-algebra $\mathbb{V}$, a valuation on $\mathbb{V}$ is a function $v : \mathsf{Prop} \to \mathbb{V}^+$. As usual, $v$ can be extended to a homomorphism $[\![-]\!]_v : \mathsf{Str} \to \mathbb{V}^+$. We say that $\mathbb{V}, v \models S \Rightarrow T$ if and only if $[\![S]\!]_v \subseteq [\![T]\!]_v$.

**Theorem 6.1** (Completeness). *Given any sequent $X \Rightarrow Y$ of $\mathbf{D.NL_\diamond}$, if $\mathbb{V}, v \models X \Rightarrow Y$ for every modal $\mathbb{K}$-algebra $\mathbb{V}$ of finite dimension and any valuation $v$ on $\mathbb{V}$, then $X \Rightarrow Y$ is a provable sequent in $\mathbf{D.NL_\diamond}$.*

As discussed in Section 2.2, $\mathbf{D.NL_\diamond}$ is complete and has the finite model property with respect to modal residuated posets. Therefore, to show Theorem 6.1, it is enough to show that any finite modal residuated poset can be embedded into the modal residuated lattice of subspaces of a modal $\mathbb{K}$-algebra of finite dimension.

Let $P$ be a finite residuated poset. We will define a modal $\mathbb{K}$-algebra $\mathbb{V}$ and a $\mathbf{D.NL_\diamond}$-morphism $h : P \to \mathcal{S}(\mathbb{V})$ which is also an order embedding.

Let $n$ be the number of elements of $P$, and let $\{p_1, \ldots, p_n\}$ be an enumeration of $P$. Let $\mathbb{V}$ be the $n^2$-dimensional vector space over $\mathbb{K}$ and let $\{e_j^i \mid 1 \leq i, j \leq n\}$ be a base. Let $h : P \to \mathbb{V}$ be defined as

$$h(p_k) = [e_j^m \mid 1 \leq j \leq n \ \ \& \ \ p_m \leq p_k].$$

We define $\star : \mathbb{V} \times \mathbb{V} \to \mathbb{V}$ on the base as follows: For every $p_k \in \mathcal{P}$ take an surjective map

$$\nu_k : n \times n \to \{e_j^m \mid 1 \leq j \leq n \ \ \& \ \ p_m \leq p_k\}$$

such that $\nu_k(m, m) = e_m^k$. Define $e_m^k \star e_r^\ell = \nu_t(m, r)$, where $p_t = p_k \otimes p_\ell$. This function uniquely extends to a bilinear map and compatible with the scalar product.

We define the relation $R \subseteq \mathbb{V} \times \mathbb{V}$ as follows $0R0$ and

$$\sum_{1 \leq i \leq d} \sum_{0 \leq j \leq d_i} \alpha_{\ell_j^i}^{k_j^i} e_{\ell_j^i}^{k_j^i} R \sum_{1 \leq i \leq d} \beta_{j_i}^{m_i} e_{j_i}^{m_i}$$

where $\alpha_{\ell_j^i}^{k_j^i}, \beta_{j_i}^{m_i} \in \mathbb{K}$, $\beta_{j_i}^{m_i} \neq 0$, $p_{k_j^i} \leq \diamond p_{m_i}$ and if $m_i = m_k$ then $j_i \neq j_k$ for $1 \leq i, k \leq d$. It is immediate that $R$ satisfies the properties of Definition 5.1.

The lemma below shows that $h$ is indeed a $\mathbf{D.NL_\diamond}$-morphism which is also an order embedding.

**Lemma 6.2.** *The following are true for the poset $P$ and $h$ as above.*

1. $p \leq q$ if and only if $h(p) \subseteq h(q)$;

2. $h(p_m \otimes p_k) = h(p_m) \otimes h(p_k)$;

3. $h(p_m \backslash p_k) = h(p_m) \backslash h(p_k)$;

4. $h(p_m / p_k) = h(p_m) / h(p_k)$;

5. $h(\Diamond p_k) = \Diamond h(p_k)$.

6. $h(\blacksquare p_k) = \blacksquare h(p_k)$.

*Proof.* 1. Assume that $p \leq q$. Let $\sum_{i,j} \alpha_j^i e_j^i$ an element of $h(p)$ where $p_i \leq p$. Then by assumption $p_i \leq q$, and therefore $\sum_{i,j} \alpha_j^i e_j^i \in h(q)$. For the other direction, assume that $p_m = p \not\leq q$, then $e_1^m \notin h(q)$, since each $e_j^i$ is independent from the rest.

2. Let $u \in h(p_m \otimes p_k)$ that is, $u = \sum_{i,j} \alpha_j^i e_j^i$ where $p_i \leq p_m \otimes p_k = p_\ell$. Since $\nu_\ell$ is surjective there is $(z_j^i, x_j^i)$ such that $\nu_\ell(z_j^i, x_j^i) = e_j^i$. By definition $e_{z_j^i}^m \star e_{x_j^i}^k = e_j^i$. Since $e_{z_j^i}^m \in h(p_m)$ and $e_{x_j^i}^k \in h(p_k)$ for each $i,j$, we have that

$$h(p_m) \otimes h(p_k) \ni \sum_{i,j} \alpha_j^i (e_{z_j^i}^m \star e_{x_j^i}^k) = \sum_{i,j} \alpha_j^i e_j^i = u.$$

Conversely let $u \in e(p_m) \otimes e(p_k)$, i.e. $u = \sum_{i,j} \alpha_j^i (e_{m_j}^{m_i} \star e_{k_j}^{k_i})$ where $p_{m_i} \leq p_m$ and $p_{k_i} \leq p_k$. Then $p_{m_i} \otimes p_{k_i} \leq p_m \otimes p_k$. Then, since $e_{m_j}^{m_i} \star e_{k_j}^{k_i} \in h(p_{m_i} \otimes p_{k_i})$, we have $e_{m_j}^{m_i} \star e_{k_j}^{k_i} \in h(p_m \otimes p_k)$ for each $i$, so $u \in h(p_m \otimes p_k)$.

3. Let $u \in h(p_m \backslash p_k)$. Then $u = \sum_{i,j} \alpha_j^i e_j^i$ where $p_i \leq p_m \backslash p_k$. By adjunction this means that $p_m \otimes p_i \leq p_k$. Pick $\sum_{i',j'} \beta_{j'}^{i'} e_{j'}^{i'} \in h(p_m)$, i.e. $p_{i'} \leq p_m$. Notice by monotonicity $p_{i'} \otimes p_i \leq p_k$. Now

$$(\sum_{i',j'} \beta_{j'}^{i'} e_{j'}^{i'}) \star (\sum_{i,j} \alpha_j^i e_j^i) = \sum_{i,i',j,j'} \beta_{j'}^{i'} \alpha_j^i (e_{j'}^{i'} \star e_j^i).$$

Each of the components are by definition in $h(p_{i'} \otimes p_i)$, and by monotonicity in $h(p_k)$. So for every $w \in h(p_m)$, $w \star u \in h(p_k)$. Therefore $u \in h(p_m) \backslash h(p_k)$.

Conversely, let $u = \sum_{i,j} \alpha_j^i e_j^i \in h(p_m) \backslash h(p_k)$. Then for every $w \in h(p_m)$, $w \star u \in h(p_k)$. In particular for $w = \sum_j e_j^m$,

$$(\sum_j e_j^m) \star (\sum_{i,j} \alpha_j^i e_j^i) \in h(p_k)$$

. Since $\star$ is bilinear and every element has a unique representation given a base, each $e_j^m \star e_j^i \in h(p_k)$. Let $p_r = p_m \otimes p_i$. By definition of $\nu_r$, $e_j^m \star e_j^i = e_j^r \in h(p_k)$ and therefore $p_m \otimes p_i \leq p_k$. That is $p_i \leq p_m \backslash p_k$, i.e. $e_j^i \in h(p_m \backslash p_k)$ for each $j$. Therefore $u \in h(p_m \backslash p_k)$.

4. The proof is the same as item 3.

5. Let $u \in h(\Diamond p_k)$, i.e., $u = \sum_i \alpha_{j_i}^{m_i} e_{j_i}^{m_i}$ where $p_{m_i} \leq \Diamond p_k$. Since $e_{j_i}^{m_i} R e_1^k$ for each $i$, it follows that $e_{j_i}^{m_i} \in R^{-1}[h(p_k)]$, for each $i$ and hence $u \in \Diamond h(p_k)$.

Conversely let $u \in \Diamond h(p_k)$, i.e. $u \in R^{-1}[h(p_k)]$. By definition of $R$ and the monotonicity of $\Diamond$ it follows that $uRe_1^k$. So $u = \sum_i \alpha_{j_i}^{m_i} e_{j_i}^{m_i}$ where $p_{m_i} \leq \Diamond p_k$, i.e. $u \in h(\Diamond p_k)$.

6. Let $u \in h(\blacksquare p_k)$. Then $u = \sum_i \beta_{j_i}^{m_i} e_{j_i}^{m_i}$ where $p_{m_i} \leq \blacksquare p_k$. By adjunction this means that $\Diamond p_{m_i} \leq p_k$. Let $vRu$ then $v = \sum_i \sum_{0 \leq j \leq n_i} \alpha_{r_j}^{\ell_j^i} e_{r_j}^{\ell_j^i}$ where $p_{\ell_j^i} \leq \Diamond p_{m_i}$. Then $p_{\ell_j^i} \leq p_k$ and therefore $v \in h(p_k)$. Hence $u \in \blacksquare h(p_k)$.

Conversely, let $u = \sum_i \beta_{j_i}^{m_i} e_{j_i}^{m_i} \in \blacksquare h(p_k)$, i.e. $v \in h(p_k)$ for every $v$ such that $vRu$. Notice that $\sum_i e_{j_i}^{\ell_i} Ru$ where $p_{\ell_i} = \Diamond p_{m_i}$. Since $v \in h(p_k)$ it follows that $\Diamond p_{m_i} \leq p_k$ and by adjunction $p_{m_i} \leq \blacksquare p_k$. Then $e_{j_i}^{m_i} \in h(\blacksquare p_k)$, for every $i$ and therefore $u \in h(\blacksquare p_k)$. □

**Remark 6.3.** *In the proof above the finiteness of $P$ was used only to guarantee the dimension of $\mathbb{V}$ to be finite. The same proof holds for an arbitrary modal residuated poset $P$ with a modal $\mathbb{K}$-algebra of dimension $|P \times P|$. That is, every modal residuated poset, and in particular the Lindenbaum-Tarski algebra of $\mathbf{D.NL_\Diamond}$, can be embedded into the lattice of subspaces of some modal $\mathbb{K}$-algebra.*

**Remark 6.4.** *In the proof of Theorem 6.1, we showed that in fact $h$ embeds $P$ into the subalgebra $\{[e_i^j \mid (i,j) \in S] \mid S \subseteq n \times n\}$ which is a Boolean subalgebra of $\mathbb{V}^+$. This is analogous to Buszkowski's proof (see e.g. [3]) that generalized Lambek calculus is complete with respect to algebraic models based on powerset algebras.*

# 7 Conclusions and further directions

**Our contributions.** In this paper we have taken a duality-theoretic perspective on vector space semantics of the basic modal Lambek calculus and some of its analytic extensions. In a slogan, we have regarded vector spaces (more specifically, modal $\mathbb{K}$-algebras) as Kripke frames. This perspective has allowed to transfer a number of results pertaining to the theory of modal logic to the vector space semantics. Our main contributions are the proof of completeness of the basic modal Lambek calculus **D.NL**$_\diamond$ with respect to the semantics given by the modal $\mathbb{K}$-algebras and a number of ensuing Sahlqvist correspondence results.

**Correspondence and completeness.** In the standard Kripke semantics setting, the completeness of the basic logic and canonicity via correspondence immediately implies that any axiomatic extension of the basic logic with Sahlqvist-type axioms is complete with respect to the elementary class of relational structures defined by the first order correspondents of its axioms. We plan to extend this result to the vector space semantics.

**Adding lattice connectives.** Another direction we plan to pursue consists in extending the present completeness result to the full Lambek calculus signature. Towards this goal, the representation results of [23, 22, 12], which embeds each complemented modular Arguesian lattice into the lattice of subspaces of a vector space (over a division ring), is likely to be particularly relevant.

**Finite vector spaces.** We plan to refine our results so as to give upper bounds on the dimensions of possible witnesses of non derivable sequents.

**Acknowledgements.** We would like to thank Peter Jipsen for numerous observations and suggestions that have substantially improved this paper. We would also like to thank the two anonymous referee for insightful remarks and suggestions.

# References

[1] M. Baroni, R. Bernardi, and R. Zamparelli. Frege in space: a program for compositional distributional semantics. *Linguistic Issues in Language Technology*, 9(241–346), 2014.

[2] W. Buszkowski. Interpolation and FEP for logics of residuated algebras. *Logic Journal of IGPL*, 19:437–454, 2011.

[3] W. Buszkowski. On involutive nonassociative Lambek calculus. *Journal of Logic, Language and Information*, 28(2):157–181, 2019.

[4] J. Chen, G. Greco, A. Palmigiano, and A. Tzimoulis. Non normal logics: Semantic analysis and proof theory. In d. Q. R. Iemhoff R., Moortgat M., editor, *Logic, Language, Information, and Computation, WoLLIC 2019*, volume 11541 of *LNCS*, pages 99–118. Springer, Berlin, Heidelberg, 2019. ArXiv:1903.04868.

[5] B. Coecke, E. Grefenstette, and M. Sadrzadeh. Lambek vs. Lambek: Functorial vector space semantics and string diagrams for Lambek calculus. *Annals of Pure and Applied Logic*, 164(11):1079–1100, 2013.

[6] B. Coecke, M. Sadrzadeh, and S. Clark. Mathematical foundations for a compositional distributional model of meaning. ArXiv:1003.4394, 2010.

[7] W. Conradie, S. Ghilardi, and A. Palmigiano. Unified Correspondence. In A. Baltag and S. Smets, editors, *Johan van Benthem on Logic and Information Dynamics*, volume 5 of *Outstanding Contributions to Logic*, pages 933–975. Springer International Publishing, 2014.

[8] W. Conradie and A. Palmigiano. Algorithmic correspondence and canonicity for non-distributive logics. *Annals of Pure and Applied Logic*, 170(9):923–974, 2019.

[9] W. Conradie, A. Palmigiano, and A. Tzimoulis. Goldblatt-Thomason for LE-logics. Submitted, 2018. ArXiv:1809.08225.

[10] J. H. Conway and D. Smith. *On Quaternions and Octonions: Their Geometry, Arithmetic, and Symmetry*. AK Peters/CRC Press, 2003.

[11] B. A. Davey and H. A. Priestley. *Introduction to lattices and order*. Cambridge university press, 2002.

[12] O. Frink. Complemented modular lattices and projective spaces of infinite dimension. *Transactions of the American Mathematical Society*, 60(3):452–467, 1946.

[13] S. Frittella, G. Greco, A. Kurz, A. Palmigiano, and V. Sikimić. A multi-type display calculus for dynamic epistemic logic. *Journal of Logic and Computation*, 26(6):2017–2065, 2016.

[14] S. Frittella, G. Greco, A. Palmigiano, and F. Yang. A multi-type calculus for inquisitive logic. In J. Väänänen, Å. Hirvonen, and R. de Queiroz, editors, *Logic, Language, Information, and Computation, WoLLIC 2016*, volume 9803 of *LNCS*, pages 215–233. Springer Berlin Heidelberg, 2016.

[15] N. Galatos, P. Jipsen, T. Kowalski, and H. Ono. *Residuated lattices: an algebraic glimpse at substructural logics*, volume 151. Elsevier, 2007.

[16] G. Greco, P. Jipsen, F. Liang, A. Palmigiano, and A. Tzimoulis. Algebraic proof theory for LE-logics. *ArXiv:1808.04642*, submitted, 2019.

[17] G. Greco, F. Liang, M. A. Moshier, and A. Palmigiano. Multi-type display calculus for semi De Morgan logic. In *Logic, Language, Information, and Computation, WoLLIC 2017*, volume 10388 of *LNCS*, pages 199–215. Springer Berlin Heidelberg, 2017.

[18] G. Greco, F. Liang, A. Palmigiano, and U. Rivieccio. Bilattice logic properly displayed. *Fuzzy Sets and Systems*, 363:138–155, 2019.

[19] G. Greco, M. Ma, A. Palmigiano, A. Tzimoulis, and Z. Zhao. Unified correspondence as a proof-theoretic tool. *Journal of Logic and Computation*, 28(7):1367–1442, 2016.

[20] G. Greco and A. Palmigiano. Lattice logic properly displayed. In *Logic, Language, Information, and Computation, WoLLIC 2017*, volume 10388 of *LNCS*, pages 153–169. Springer Berlin Heidelberg, 2017.

[21] G. Greco and A. Palmigiano. Linear logic properly displayed. Submitted, ArXiv:1611.04181.

[22] B. Jónsson. On the representation of lattices. *Mathematica Scandinavica*, 1:193–206, 1953.

[23] B. Jónsson. Modular lattices and Desargues' theorem. *Marhematica Scandinavica*, 2:295–314, 1955.

[24] Y. Kubota and R. Levine. Gapping as like-category coordination. In D. Béchet and A. J. Dikovsky, editors, *Logical Aspects of Computational Linguistics - 7th International Conference, LACL 2012, Nantes, France, July 2-4, 2012. Proceedings*, volume 7351 of *Lecture Notes in Computer Science*, pages 135–150. Springer, 2012.

[25] J. Lambek. The mathematics of sentence structure. *The American Mathematical Monthly*, 65(3):154–170, 1958.

[26] J. Lambek. On the calculus of syntactic types. In R. Jakobson, editor, *Structure of Language and its Mathematical Aspects*, volume XII of *Proceedings of Symposia in Applied Mathematics*, pages 166–178. American Mathematical Society, 1961.

[27] S. Lang. *Linear Algebra*. Springer Undergraduate Texts in Mathematics and Technology. Springer, 1987.

[28] M. Moortgat. Multimodal linguistic inference. *Journal of Logic, Language and Information*, 5(3-4):349–385, 1996.

[29] M. Moortgat and G. Wijnholds. Lexical and derivational meaning in vector-based models of relativisation. In A. Cremers, T. van Gessel, and F. Roelofsen, editors, *Proceedings of the 21st Amsterdam Colloquium*, pages 55–64. ILLC, University of Amsterdam, 2017.

[30] G. Morrill, O. Valentín, and M. Fadda. The displacement calculus. *Journal of Logic, Language and Information*, 20(1):1–48, 2011.

[31] M. Sadrzadeh, S. Clark, and B. Coecke. The Frobenius anatomy of word meanings I: Subject and object relative pronouns. *Journal of Logic and Computation*, pages 1293–1317, 2013.

[32] H. Wansing. *Displaying Modal Logic*. Kluwer, 1998.

www.ingramcontent.com/pod-product-compliance
Lightning Source LLC
Chambersburg PA
CBHW080449170426
43196CB00016B/2738